# AN ELEMENTARY TREATISE ON THE LUNAR THEORY

# AN ELEMENTARY TREATISE ON THE LUNAR THEORY:

WITH

A BRIEF SKETCH OF THE PROBLEM BEFORE NEWTON

HUGH GODFRAY

GYAN

*Published by*

GYAN

Gyan Publishing House
5, Ansari Road
Daryaganj, New Delhi-110002
Phone: 011-47034999, 9811692060
E-mail: books@gyanbooks.com

*Distribution Network*
gyanbooks.com
India, USA, Canada, UK, Australia



ISBN: 978-93-7014-586-3 (PB)

First Published, 1871

2nd Impression 2025

*Printed at:* Gyan Press, Delhi.



**AN ELEMENTARY TREATISE ON THE LUNAR THEORY**

Author: HUGH GODFRAY

# AN ELEMENTARY TREATISE ON THE LUNAR THEORY.

AN

# ELEMENTARY TREATISE

ON THE

# LUNAR THEORY,

WITH

A BRIEF SKETCH OF THE HISTORY OF THE PROBLEM BEFORE NEWTON.

BY

HUGH GODFRAY, M.A.

OF ST. JOHN'S COLLEGE, CAMBRIDGE;
LATE MATHEMATICAL LECTURER AT PEMBROKE COLLEGE.

FOURTH EDITION.

London and New York:
MACMILLAN AND CO.
1885.

## PREFACE TO THE FIRST EDITION.

Of all the celestial bodies whose motions have formed the subject of the investigations of astronomers, the Moon has always been regarded as that which presents the greatest difficulties, on account of the number of inequalities to which it is subject; but the frequent and important applications of the results render the Lunar Problem one of the highest interest, and we find that it has occupied the attention of the most celebrated astronomers from the earliest times.

Newton's discovery of Universal Gravitation, suggested, it is supposed, by a rough consideration of the motions of the Moon, led him naturally to examine its application to a more severe explanation of her disturbances; and his Eleventh Section is the first attempt at a theoretical investigation of the Lunar inequalities. The results he obtained were found to agree very nearly with those determined by observation, and afforded a remarkable confirmation of the truth of his great principle; but the geometrical methods which he had adopted seem inadequate to so complicated a theory, and recourse has been had to analysis for a complete determination of the disturbances, and for a knowledge of the true orbit.

The following pages will, it is hoped, form a proper introduction to more recondite works on the subject: the difficulties which a person entering upon this study is most likely to stumble at have been dwelt upon at considerable length, and though different methods of investigation have been employed by different astronomers, the difficulties met with are nearly the same, and the principle of successive approximation is common to all. In the present work, the approximation is carried to the second order of small quantities, and this, though far from giving accurate values, is amply sufficient for the elucidation of the method.

The differences in the analytical solutions arise from the various ways in which the position of the moon may be indicated by altering the system of coordinates to which it is referred, or again, in the same system, by choosing different quantities for independent variables.

D'Alembert and Clairaut chose for coordinates the projection of the radius vector on the plane of the ecliptic and the longitude of this projection. To form the differential equations, the true longitude was taken for independent variable.

To determine the latitude, they, by analogy to Newton's method, employed the differential variations of the motion of the node and of the inclination of the orbit.

Laplace, Damoiseau, Plana, and also Herschel and Airy, in their more elementary works, have found it more convenient to express the variations of the latitude directly, by an equation of the same form as that of the radius vector.

Lubbock and Pontécoulant, taking the same coordinates of the Moon's position, make the time the independent variable; and when it is desired to carry the approximation to a high order, this method offers the advantage of not requiring any reversion of series.

Poisson proposed the method used in the planetary theory, that is, to determine the variation in the elements of the Moon's orbit, and thence to conclude the corresponding variations of the radius vector, the longitude, and the latitude.

The selection of the method followed in the present work, which is the same as that of Airy, Herschel, &c., was made on account of its simplicity; moreover, it is the method which has obtained sanction in this University, and it is hoped that it may prove of service to the student in his reading for the examination for Honours. In furtherance of this object, one of the chapters (the sixth) contains the physical intrepretation of the various important terms in the radius vector, latitude, and longitude.*

The seventh chapter, or appendix, contains some of the most interesting results in the terms of the higher orders, among which will be found the values of $c$ and $g$ completely obtained to the third order.

The last chapter is a brief historical sketch of the Lunar Problem before Newton, containing an account of the discoveries of the several inequalities and of the methods by which they were represented, those only being mentioned which, as the theory has since verified, were real onward steps. The perusal of this chapter will shew to what extent we are indebted to our great philosopher; at the same time we cannot fail being impressed with revercnce for the

* See the Report of the "Board of Mathematical Studies" for 1850.

genius and preservance of the men who preceded him, and whose elaborate and multiplied hypotheses were in some measure necessary to the discovery of his simple and single law.

I take this opportunity of acknowledging my obligations to several friends, whose valuable suggestions have added to the utility of the work.

HUGH GODFRAY.

*St. John's College, Cambridge,*
*April*, 1853.

---

## PREFACE TO THE THIRD EDITION.

In the present edition several important alterations have been made, many of the parts have been entirely re-written, and a general re-arrangement of the whole subject has taken place. In Chap. IV. Sect. 3, the necessity for the computation of the elements in the order $s$, $u$, $\theta$ is shewn. In the *Quarterly Journal*, Mr. Walton pointed out the logical necessity for having the value of $u$ completely correct to the second order before proceeding to find $\theta$, although the result is the same as when the preparation for both values is carried on simultaneously. I have here proved that there is the same necessity for making the value of $s$ precede that of $u$, although, again, the change has no effect on the result.

The term of the third order "Parallactic Inequality" will be found completely investigated in the appendix; and at the end of the volume a selection of questions from University and College Examination Papers has been introduced.

*August*, 1871.

# CONTENTS.

## CHAPTER I.

### INTRODUCTION.



## CHAPTER II.

### MOTION RELATIVE TO THE EARTH.



## CHAPTER III.

### DIFFERENTIAL EQUATIONS.



## CHAPTER IV.

### INTEGRATION OF THE DIFFERENTIAL EQUATIONS.

## CHAPTER V.

### NUMERICAL VALUES OF THE COEFFICIENTS.



## CHAPTER VI.

### PHYSICAL INTERPRETATION.



## CHAPTER VII.

### APPENDIX.

## CHAPTER VIII.

### HISTORY OF THE LUNAR PROBLEM BEFORE NEWTON.



## SELECTION OF EXAMINATION QUESTIONS

# LUNAR THEORY.

## CHAPTER I.

### INTRODUCTION.

BEFORE proceeding to the consideration of the moon's motion, it will be desirable to say a few words on the law of attractions, and on the peculiar circumstances which enable us to simplify the present investigation.

1. The law of universal gravitation, as laid down by Newton, is that "*Every particle in the universe attracts every other particle, with a force varying directly as the mass of the attracting particle and inversely as the square of the distance between them.*"

The truth of this law cannot be established by abstract reasoning; but as it is found that the motions of the heavenly bodies, calculated on the assumption of its truth, agree more and more closely with the observed motions the more strictly our calculations are performed, we have every reason to consider the law as an established truth, and to attribute any slight discrepancy between the results of calculation and observation to instrumental errors, to an incomplete analysis, or to our ignorance of the existence of some of the disturbing causes.

Of the last source of error there is a remarkable example connected with the discovery of the planet Neptune, which became known to us, as one of the bodies of our system,* solely by means of the perturbations it produced in the calculated places of the planet Uranus. These perturbations were too great to be attributed wholly to errors of instruments or of calculation; and therefore, either the law of universal gravitation ceased to hold for a body so remote as Uranus, or else some unknown cause was disturbing the path of the planet. The first supposition was too repugnant to an astronomer of the nineteenth century to be entertained until all others had failed,—and the second supposition led to the detection of Neptune. The distinguished names of Adams and Le Verrier will be for ever connected with the history of this planet, and their solution of the difficult inverse problem:—"Given the perturbations caused by an unknown planet, determine, on the assumption of the truth of Newton's law, the orbit and position of this disturbing body,"—will always be considered one of the triumphs of Mathematical Science.

Another remarkable instance of the vindication of Newton's law, where it seemed at first to be at fault, will be found mentioned in Chap. VI. in connection with the motion of the Lunar Apogee. The error was owing to an incomplete analysis; but, when the calculations were carried out more fully, the strongest confirmation of the law was afforded by the exact agreement of the results with observation.

We shall proceed to apply the law of gravitation to investigate the circumstances of the moon's motion; and then see how it will enable us to assign her position at any time when sufficient data have been obtained by observation.

2. The problem would be one of extreme, if not insurmountable difficulty, if we had to take into account simul-

* It had been seen by Dr. Lamont at Munich, one year before its being known to be a planet. "Solar System, by J. R. Hind."

taneonsly the actions of the earth, sun, planets, &c., on the moon; but fortunately the planets are so small or so distant that their action may be neglected—at any rate to the order of our present approximation—and the attraction of the earth, on account of its proximity, is very much greater than the *disturbing** action of the sun notwithstanding his enormous mass. We may therefore treat the question as that of one body, the moon, revolving about another body, the earth, and continually disturbed by a third body, the sun. This is the celebrated problem of *Three Bodies.*

Stated in this general form its exact solution has hitherto defied the powers of the analyst. The disturbing action of the sun can be expressed without difficulty in terms of the masses and distances, but the integration of the differential equations cannot be effected. If however, instead of retaining the whole disturbing force of the sun, which, as we have said, is small compared with the earth's action, we expand its expression in a series and neglect all very small terms, it is then found possible to obtain a solution.

This breaking up of the expression for the sun's action is obviously equivalent to a breaking up of the sun's force; and if we afterwards wish to take into account the very small forces which have been thus neglected, or the small disturbances due to the action of the planets, the following principle will shew that we may do so by considering their effects separately,—and that the algebraical sum of all the disturbances so obtained will, to an order of approximation far beyond that which we contemplate, be the same as the disturbance due to their simultaneous action. We shall find

---

* Since the sun attracts both the earth and moon, it is clear that its effects on the moon's motion relatively to the earth, or the disturbing force, will not be the same as the absolute force on either body. The absolute force of the sun on the moon is more than double that of the earth (Art. 99), but the relative force does not exceed $\frac{1}{108}$ of the earth's action (Art. 25).

an application of this principle in the investigation of the parallactic inequality.

*Principle of Superposition of Small Disturbances.*

3. Let a particle be moving under the action of any number of independent forces, some of which are very small, and let $A$ be the position of the particle at any instant. Let two of these small forces $m_1$, $m_2$ be omitted, and suppose the path of the particle under the action of the remaining forces to be $AP$ in any given time.

Let $AP_1$ be the path which would have been described in the same time if $m_1$ also had acted; $AP_1$ differing very slightly from $AP$, the disturbance being $PP_1$.

Similarly, if $m_2$ instead of $m_1$ had acted, suppose $AP_2$ to represent the disturbed path, $PP_2$ being the disturbance, ($AP$, $AP_1$, $AP_2$ are not necessarily in the same plane, nor even plane curves).

Lastly, let $AQ$ be the actual path of the body when both $m_1$ and $m_2$ act. Join $P_1Q$.

The two disturbances $PP_2$ and $P_1Q$, being due to the action of the same small force $m_2$ on the slightly different paths $AP$ and $AP_1$, must be very nearly equal both in magnitude and in direction. What difference exists between them must be a very small fraction of either disturbance, and may be neglected compared with the original path. We may, in fact, expect that this difference which is the disturbance due to $m_1$ of a disturbance due to $m_2$ will be of an order compounded of the orders of the two. Thus, if the disturbance due to $m_1$ and $m_2$ be respectively of the second and fourth orders, the difference between $P_1Q$ and $PP_2$ would probably be of the sixth order compared with $AP$. Therefore $P_1Q$ may be considered parallel and equal to $PP_2$.

Hence the projection of the whole disturbance $PQ$ on any straight line, being equal to the algebraical sum of the projections of $PP_1$ and $P_1Q$, will be equal to the algebraical sum of the projections of the separate disturbances $PP_1$, $PP_2$.

Next, let there be three small disturbing forces $m_1$, $m_2$, $m_3$. We may consider the joint action of the two $m_2$, $m_3$, as one small disturbing force; therefore, by what precedes, the total disturbance along any axis will be the snm of the separate disturbances of $m_1$ and of the system $m_2$, $m_3$; but this last is the sum of the separate disturbances of $m_2$ and $m_3$: therefore the whole disturbance equals the sum of the three separate disturbances.

4. This reasoning can evidently be extended to any number of forces; and if $x$, $y$, $z$ be the coordinates of the disturbed particle, $\phi(x, y, z)$ any function of $x$, $y$, $z$; the disturbance produced in $\phi(x, y, z)$ will be

$$\delta\phi(x, y, z) = \frac{d\phi}{dx}\delta x + \frac{d\phi}{dy}\delta y + \frac{d\phi}{dz}\delta z, \text{ omitting } (\delta x)^2, \text{ \&c.},$$

where $\delta x = \delta x_1 + \delta x_2 + \ldots =$ sum of disturbances along axis of $x$ due to separate forces,

$\delta y = \delta y_1 + \delta y_2 + \ldots =$ ........................ along axis of $y$,

$\delta z = \delta z_1 + \delta z_2 + \ldots =$ ........................ along axis of $z$;

$$\text{therefore } \delta\phi(x, y, z) = \frac{d\phi}{dx}\delta x_1 + \frac{d\phi}{dy}\delta y_1 + \frac{d\phi}{dz}\delta z_1$$

$$+ \frac{d\phi}{dx}\delta x_2 + \frac{d\phi}{dy}\delta y_2 + \frac{d\phi}{dz}\delta z_2 + \text{\&c.}$$

$$= \delta_1\phi(x, y, z) + \delta_2\phi(x, y, z) + \text{\&c.},$$

or total disturbance equals sum of separate disturbances, which establishes the principle.

Since $\phi(x, y, z)$ may be the radius vector, the latitude or the longitude of the disturbed body, it follows that the total disturbance in any of these elements is the sum of the partial disturbances.

5. But we must still prove another proposition, without which the problem, though so far simplified, would scarcely be less complicated than in its most general form.

Newton's law refers to *particles*, whereas the sun, earth, and moon are *large* nearly spherical bodies, and it becomes necessary to examine the mutual action of such bodies. Now, it happens that, with this law of force, the attraction of one sphere on another can be correctly obtained, and leaves the question in exactly the same state as if they were particles. (*Princip.* lib. I. prop. 75.)

*Attractions of Spherical Bodies.*

6. Let $P$ be a particle situated at a distance $OP = a$ from the centre of a sphere of uniform density $\rho$ and radius $c$. The particle being without the sphere, $a > c$.

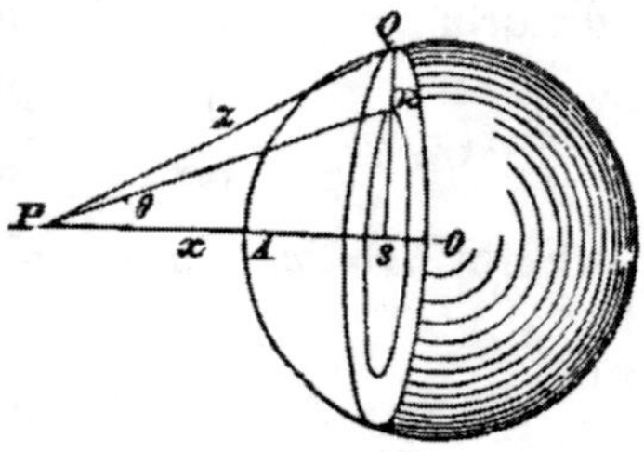


Let the whole sphere be divided into circular laminæ by planes perpendicular to $OP$. Let $SQ$ be one of these. $PS = x$, $PQ = z$, and thickness of lamina $= \delta x$.

Next, let this lamina be divided into concentric rings. Let $RS = r$ be the radius of one of these rings and $\delta r$ its breadth, $\angle RPO = \theta$;

therefore
$$r = x \tan\theta,$$
$$\delta r = x \sec^2\theta \, . \, \delta\theta.$$

The attraction of an element $R$ of this ring on the particle $P$ will be $\dfrac{\text{mass of element}}{PR^2}$ along $PR$, and the resolved part of this along $PO$ will be $\dfrac{\text{mass of element}}{x^2 \sec^2\theta} \cos\theta$.

But the resultant attraction of the whole ring will clearly be the sum of the resolved parts along $PO$ of the attractions of its constituent elements; therefore,

$$\text{attraction of ring} = \frac{2\pi\rho r\,.\,\delta r\,.\,\delta x}{x^2 \sec^2\theta}\cos\theta = 2\pi\rho\sin\theta\delta x\,.\,\delta\theta\,;$$

therefore, attraction of whole lamina $SQ = 2\pi\rho\delta x\int_0^{\cos^{-1}\frac{x}{z}}\sin\theta.d\theta$

$$= 2\pi\rho\delta x\left(1 - \frac{x}{z}\right).$$

Again, $z^2 = x^2 + c^2 - (a - x)^2 = 2ax - (a^2 - c^2)$;

therefore $z\delta z = a\delta x$,

and attraction of lamina $= 2\pi\rho\left(\frac{z\delta z}{a} - \frac{z^2 + a^2 - c^2}{2a^2}\delta z\right)$;

$\therefore$ attraction of whole sphere $= 2\pi\rho\left\{\frac{z^2}{2a} - \frac{z^3}{6a^2} - \frac{(a^2 - c^2)z}{2a^2}\right\}$

(from $z = a - c$ to $z = a + c$)

$$= \frac{4\pi\rho c^3}{3a^2}$$

$$= \frac{M}{a^2},$$

where $M$ = mass of sphere $= \frac{4\pi\rho c}{3}$.

Hence, *the attraction of the whole sphere is precisely the same as if the whole mass were condensed into its centre.*

COR. 1. The attraction of a shell (radius $c$ and thickness $\delta c$) will be obtained from the preceding expression by differentiating it with respect to $c$, and is

$$\text{attraction of shell} = \frac{4\pi\rho c^2.\delta c}{a^2} = \frac{\text{mass of shell}}{a^2},$$

the same as if the mass were collected at its centre.

COR. 2. Therefore, *the attraction of a* heterogeneous *sphere on an external particle will be the same as if the whole mass were condensed into its centre, provided the density be the same at all points equally distant from the centre*, for then the whole sphere may be considered as the aggregate of an infinite number of uniform shells.

7. Let us now consider the case of one sphere attracting another. Suppose $P$ in the preceding article to be an elementary particle of a sphere $M'$, whose centre $O'$ suppose at a distance $a$ from $O$. Then, since action and reaction are equal and opposite, $P$ will attract the whole sphere $M$ just as it would do a particle of mass $M$ placed at $O$. The same is true of all the elementary particles which compose the sphere $M'$, therefore the sphere $M'$ will attract the sphere $M$ as if the whole mass of the latter were condensed into its centre $O$; but the attraction of the sphere $M'$ on a particle $O$ is the same as if the attracting sphere were condensed into its centre $O'$; therefore,

*Two spheres attract one another as if the whole matter of each sphere were collected at its centre.*

8. This remarkable result, which, as may be shewn, holds only when the law of attraction is that of the inverse square of the distance, or that of the direct distance, or a combination of these by addition or subtraction, reduces the problem of the sun, earth, and moon to that of three particles. The slight error due to the bodies not being perfect spheres is here neglected, being of an order higher than that to which we intend to carry the present investigation: this error, however, though very small, is appreciable, and when a nearer approximation is required, it becomes necessary to have regard to this circumstance. (See Appendix, Art. 109).

# CHAPTER II.

## MOTION RELATIVE TO THE EARTH.

9. When a number of particles are in motion under their mutual attractions or other forces, and the motion relatively to one of them is required, we must bring that one to rest and then keep it at rest, without altering the relative motions of the others with respect to it.

Now, first, the chosen particle will be brought to rest by giving it at any instant a velocity equal and opposite to that which it has at that instant; secondly, it will be kept at rest by applying to it accelerating forces equal and opposite to those which act upon it.

Therefore give the same velocity and apply the same accelerating forces to all the bodies of the system, and their absolute motions about the chosen body, which is now at rest, will be the same as their relative motions previously.

### *Problem of Two Bodies.*

10. As the sun disturbs the moon's motion with respect to the earth, it is important to know what that motion would have been if no disturbance had existed, or generally:—

*Two bodies attracting one another with forces varying directly as the mass and inversely as the square of the distance, to determine the orbit of one relatively to the other.*

Let $M$, $M'$ be the masses of the bodies, $M'$ being the body whose motion relatively to $M$ is required, $r$ the distance between them at any time $t$, and $\theta$ the angle between $r$ and some fixed prime radius.

The accelerating force of $M$ on $M'$ equals $\frac{M}{r^2}$ acting towards $M$, while that of $M'$ on $M$ equals $\frac{M'}{r^2}$ in the opposite direction. Therefore, by the principle above stated, we must apply to both $M$ and $M'$ accelerating forces equal and opposite to this latter force, and $M'$ will move about $M$ fixed, the accelerating force on $M'$ being $\frac{M+M'}{r^2} = \mu u^2$, if $\mu = M + M'$ and $r = \frac{1}{u}$.

Hence,
$$\frac{d^2u}{d\theta^2} + u = \frac{\mu}{h^2},$$
where $h = r^2 \frac{d\theta}{dt}$ = twice the area described in a unit of time, and integrating
$$u = \frac{\mu}{h^2}\{1 + e\cos(\theta - \alpha)\},$$
$e$ and $\alpha$ being constants to be determined by the circumstances of the motion at any given time.

This is the equation to a conic section referred to its focus, the eccentricity being $e$, the semi-latus rectum $\frac{h^2}{\mu}$, and the angle made by the apse line with the prime radius $\alpha$.

By observations made on the actual path of the moon, which cannot differ very widely from its undisturbed orbit, we infer that this latter would be an ellipse with an eccentricity about $\frac{1}{20}$.

In the same manner, if the sun and earth be the two bodies considered, the relative orbit would be an ellipse with an eccentricity about $\frac{1}{60}$.

11. The angle $\theta - \alpha$ between the radius vector and the apse line is called the *true anomaly*.

If $n$ is the angular velocity of a radius vector which, moving uniformly, would accomplish its revolution in the same time as the true one, both passing through the apse at the same instant; then $nt+\epsilon-\alpha$ is called the *mean anomaly* where $\epsilon$ is a constant depending on the instant from which the time is reckoned, its value being the angle between the prime radius and the uniformly revolving one when $t=0$.

Thus, let $M\Upsilon$ be the fixed line or prime radius,

$A$ the apse,

$M'$ the moving body at time $t$,

$M\mu$ the uniformly revolving radius at same time, the direction of motion being represented by the arrow.

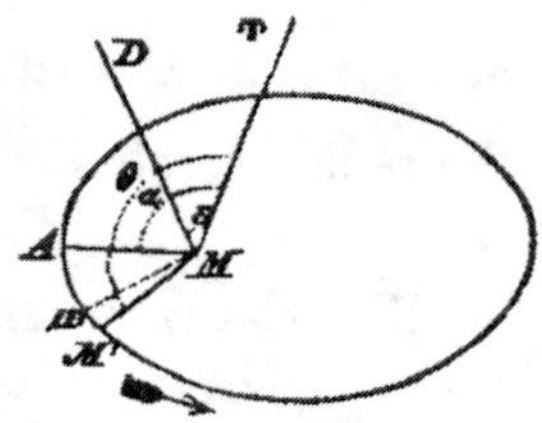

And, let $MD$ be the position of $M\mu$ when $t=0$,

then $\quad \Upsilon MD=\epsilon$ and is called the epoch,*

$DM\mu=nt$,

$\Upsilon MA=\alpha=$ longitude of the apse;

therefore, mean anomaly $=AM\mu=nt+\epsilon-\alpha$,

true anomaly $=AMM'=\Upsilon MM'-\Upsilon MA=\theta-\alpha$.

12. *To express the* mean *anomaly in terms of the* true *in a series ascending according to the powers of* e, *as far as* $e^3$.

$$n=\frac{2\pi}{\text{periodic time}}=2\pi\div\frac{2\text{ area}}{h}$$

$$=\frac{2\pi h}{2\pi ab}=\frac{h}{a^2\sqrt{(1-e^2)}};$$

* We shall see (Art. 36) that the introduction of the epoch $\epsilon$ is avoided in the Lunar Theory by a particular assumption; but in the Planetary it forms one of the important elements of the orbit.

therefore $$h = na^2(1-e^2)^{\frac{1}{2}},$$

$$\frac{dt}{d\theta} = \frac{r^2}{h} = \frac{a^2(1-e^2)^2}{h} \cdot \frac{1}{\{1+e\cos(\theta-\alpha)\}^2}$$

$$= \frac{1}{n}(1-e^2)^{\frac{3}{2}}\{1+e\cos(\theta-\alpha)\}^{-2}$$

$$= \frac{1}{n}(1-\tfrac{3}{2}e^2)\{1-2e\cos(\theta-\alpha)+3e^2\cos^2(\theta-\alpha)\}$$

$$= \frac{1}{n}\{1-2e\cos(\theta-\alpha)+\tfrac{3}{2}e^2\cos 2(\theta-\alpha)\};$$

therefore $nt+\varepsilon = \theta - 2e\sin(\theta-\alpha) + \tfrac{3}{4}e^2\sin 2(\theta-\alpha)$,

or $(nt+\varepsilon-\alpha) = (\theta-\alpha) - 2e\sin(\theta-\alpha) + \tfrac{3}{4}e^2\sin 2(\theta-\alpha)$,

the required relation.

13. *To express the* true *anomaly in terms of the* mean *to the same order of approximation.*

$$\theta-\alpha = nt+\varepsilon-\alpha+2e\sin(\theta-\alpha)-\tfrac{3}{4}e^2\sin 2(\theta-\alpha)\ldots(1);$$

$\therefore\ \theta-\alpha = nt+\varepsilon-\alpha$ first approximation.

Substituting this in the first small terms of (1), we get

$\theta-\alpha = nt+\varepsilon-\alpha+2e\sin(nt+\varepsilon-\alpha)$...a second approximation.

Substitute the second approximation in that small term of (1) which is multiplied by $e$, and the first approximation in that multiplied by $e^2$, the result will be correct to that term, and gives

$$\theta-\alpha = nt+\varepsilon-\alpha+2e\sin\{nt+\varepsilon-\alpha+2e\sin(nt+e-\alpha)\} - \tfrac{3}{4}e^2\sin 2(nt+\varepsilon-\alpha)$$

$$= nt+\varepsilon-\alpha+2e\sin(nt+\varepsilon-\alpha) + 4e^2\cos(nt+\varepsilon-\alpha)\sin(nt+\varepsilon-\alpha)-\tfrac{3}{4}e^2\sin 2(nt+\varepsilon-\alpha)$$

$$= nt+\varepsilon-\alpha+2e\sin(nt+\varepsilon-\alpha)+\tfrac{5}{4}e^2\sin 2(nt+\varepsilon-\alpha),$$

the required relation.

The development could be carried on by the same process to any power of $e$, the coefficient of $e^3$ would be found

$$\tfrac{13}{12}\sin 3\,(nt+\varepsilon-\alpha)-\tfrac{1}{4}\sin\,(nt+\varepsilon-\alpha),$$

but in what follows we shall not require anything beyond $e^2$.

*Problem of Three Bodies.*

14. In order to fix the position of the moon with respect to the centre of the earth, which, by means of the process described in Art. (9), is supposed brought to rest, we must have some determinate invariable plane passing through the earth's centre to which the motion may be referred.

If the sun and earth were the only bodies in the universe, then the plane of the elliptic orbit (in which, according to the last section, the motion would take place) would be fixed, and might be taken as the plane of reference; but, as soon as we take into account the action of the moon and planets—especially the moon—the plane ceases to be fixed, and some other plane must be found not affected by these disturbances.

Theory teaches us that such a plane exists,* but as its exact determination can only be the work of time, the

---

* See *Poinsot*, "*Théorie et détermination de l'équateur du système solaire,*" where he proves that an invariable plane exists for the solar system, that is, a plane whose position relatively to the fixed stars will always be the same, whatever changes the orbits of the planets may experience; but as its position depends on the moments of inertia of the sun, planets, and satellites, and therefore on their internal conformation, it cannot be determined *à priori*, and ages must elapse before observation can furnish sufficient data for doing so *à posteriori*.

This result Poinsot obtains on the supposition that the solar system is a free system; but it is possible, as be furthermore remarks, nay probable, that the stars exert some action upon it, it follows that this *invariable* plane may itself be variable, though the change must, according to our ideas of time and space, be indefinitely slow and small.

following theorem will supply us with a plane whose motion is extremely slow, and which may for a very long period, and to a degree of approximation far beyond that to which we shall carry our investigations, be considered as fixed and coinciding with its position at present.

In what follows the action of the planets is too small to be taken into account.

15. *The centre of gravity of the earth and moon describes relatively to the sun an orbit very nearly in one plane and elliptic;* the square of the ratio of the distances of the moon and sun from the earth being neglected.*

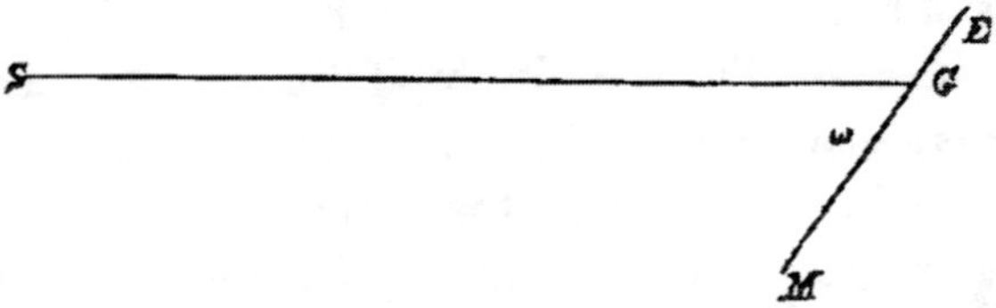


Let $S$, $E$, $M$ be the centres of the sun, earth, and moon, $G$ the centre of gravity of the last two. Now the motion of $G$ is the same as if the whole mass $E+M$ were collected there and acted on by forces equal and parallel to the moving forces which act on $E$ and $M$. The whole force on $G$ is therefore in the plane $SEM$; join $SG$.

Let $\angle SGM = \omega$, and let $m'$ be the sun's absolute force.

$$\text{Moving force on } E = \frac{m'.E}{SE^2} \text{ in } ES,$$

$$\text{moving force on } M = \frac{m'.M}{SM^2} \text{ in } MS.$$

---

* This ratio is about $\frac{1}{400}$, and, as we shall see Art. (21), such a quantity we shall consider as of the 2nd order of small quantities, and its square therefore of the 4th order. Our investigations are carried to the 2nd order only.

These applied to $G$ parallel to themselves are equivalent to

$$\left.\begin{array}{ll} & \dfrac{m'.E.GE}{SE^3} - \dfrac{m'.M.GM}{SM^3} \text{ in direction } GM, \\ \text{and} & \dfrac{m'.E.SG}{SE^3} + \dfrac{m'.M.SG}{SM^3} \text{ ............... } GS; \end{array}\right\} \begin{array}{l}\text{by the} \\ \text{triangle} \\ \text{of forces.}\end{array}$$

but
$$\frac{E}{GM} = \frac{M}{GE} = \frac{M+E}{ME},$$

whence
$$E.GE = M.GM = (M+E)\frac{GM.GE}{ME}.$$

Therefore substituting, and dividing by $M+E$, we get accelerating forces*

$$m'.\frac{GM.GE}{ME}\left(\frac{1}{SE^3} - \frac{1}{SM^3}\right) \text{ in direction } GM,$$

and
$$\frac{m'.SG}{M+E}\left(\frac{E}{SE^3} + \frac{M}{SM^3}\right) \text{ .............. } GS.$$

Again,
$$SE^2 = SG^2 + GE^2 + 2SG.GE\cos\omega$$
$$= SG^2\left(1 + \frac{2GE}{SG}\cos\omega + \frac{GE^2}{SG^2}\right);$$

$$\left.\begin{array}{ll} \text{therefore} & \dfrac{1}{SE^3} = \dfrac{1}{SG^3}\left(1 - \dfrac{3GE}{SG}\cos\omega\right) \\ \text{similarly} & \dfrac{1}{SM^3} = \dfrac{1}{SG^3}\left(1 + \dfrac{3GM}{SG}\cos\omega\right) \end{array}\right\} \begin{array}{l}\text{omitting quantities} \\ \text{to be neglected.}\end{array}$$

---

* In strictness it would be necessary, since we have brought $S$ to rest to apply to both $M$ and $E$, and therefore to $G$, accelerating forces equal and opposite to those which $E$ and $M$ themselves exert on $S$; but the mass of $S$ is so large compared with those of $E$ and $M$, that we may safely neglect these forces in this approximate determination of the path of $G$, the error being of a still higher order than that introduced by the neglect of $\left(\frac{EM}{SG}\right)^2$.

Therefore the accelerating force in the direction $GM$

$$= -\frac{3m'.GM.GE}{SG^4}\cos\omega$$

$$= -3\,(\text{accelerating force of sun on } G)\,\frac{GM}{SG}\cdot\frac{GE}{SG}\cos\omega$$

$= 0$, according to standard of approximation adopted.

And the accelerating force in $GS$

$$= \frac{m'.SG}{M+E}\left\{\frac{E+M}{SG^3} - \frac{3(E.GE - M.GM)}{SG^4}\cos\omega\right\}$$

$$= \frac{m'}{SG^2} \text{ to the same approximation.}$$

Hence, the force on $G$ is a central force tending to $S$ and varying inversely as the square of the distance; therefore the orbit of $G$ about $S$ is very approximately an ellipse with $S$ in the focus; and the plane of this ellipse is, as far as our investigations are concerned, a fixed plane.

This fixed plane is called the *plane of the ecliptic*, or simply *the ecliptic.*

16. We shall now suppose a plane through the earth's centre parallel to the ecliptic; and this, when the earth is brought to rest, will give us the fixed plane of reference we require (14); the ecliptic then making small monthly oscillations from one side to the other of our fixed plane.

17. Since $G$ describes an ellipse relatively to the sun, the sun will describe the same ellipse relatively to $G$; but, as seen from the earth, the orbit of the sun will be slightly different, as we shall now shew.

First: The sun will have a latitude, that is, will be out of the plane of reference; and this latitude will be of the same name (north or south) as that of the moon, and deducible from it when the ratio of the distances of the two bodies from the earth, and that the masses of the earth

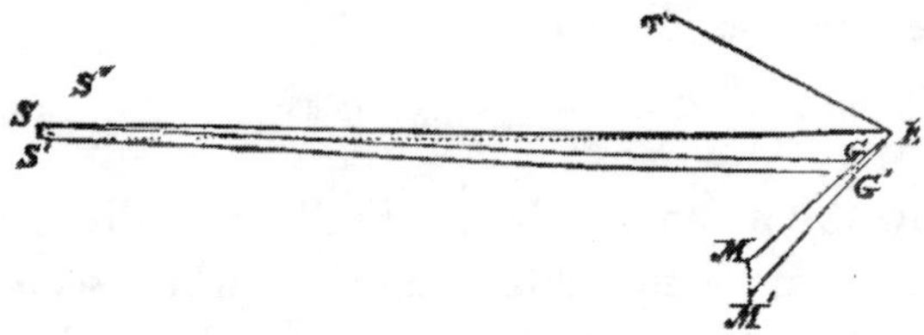


and moon are known. For if $S'EM'$ be the fixed plane through $E$, and $S'$, $G'$, $M'$ the projections of $S$, $G$, $M$, then $SES'$ is the sun's latitude, and $MEM'$ is the moon's; and the two bodies are obviously both on the same side of the plane.

$$\sin(\text{sun's lat.}) = \frac{SS'}{ES} = \frac{GG'}{ES} = \frac{EG.\sin(\text{moon's lat.})}{ES}$$

$$= \frac{M}{E+M} \cdot \frac{EM}{ES} \cdot \sin(\text{moon's lat.}).$$

Now, it is known that $M$ is about $\frac{1}{80}$ of $E$,

and $EM$......... $\frac{1}{400}$ of $ES$;

therefore $\sin(\text{sun's lat.}) = \dfrac{\sin(\text{moon's lat.})}{32400}$ nearly.

And as the moon's latitude never exceeds 5° 9′, the sun's latitude will always be less than 1″.

Secondly, as to the sun's longitude: let $E\Upsilon$ be the direction of the first point of Aries,—that is, a fixed line in our plane of reference from which the longitudes of the bodies are reckoned. Then $\Upsilon ES' = \theta'$ is the sun's longitude as seen from $E$; but, as seen from $G$, the sun's longitude would be the angle between $GS$ and a line through $G$ (not drawn in the figure) parallel to $E\Upsilon$.

The difference between these two longitudes is the angle $ES'G'$; and, if $EG'S' = \omega$,

$$\sin ES'G' = \frac{EG'}{ES'} \sin\omega = \frac{\sin\omega}{32400} \text{ approximately};$$

therefore $\sin ES'G'$ never exceeds $\frac{1}{32400}$, and $ES'G'$ is a small angle not exceeding 7″.

Also $ES' \sim S'G' < EG' < \frac{1}{32400} S'G'$.

18. Now, by assuming the longitude and distance of the sun as seen from $E$ to be the same as when seen from $G$, we commit the above small errors in the position of $S$; that is, we assume the sun to be at $S''$ instead of $S$, $S'S''$ being drawn equal and parallel to $G'E$. If our object were the determination of the sun's position, it would be necessary to take this into account; but when investigating the motion of the moon, the disturbing action of the sun will not, on account of his great distance, be appreciably altered by supposing him to be at $S''$ instead of $S$; and the errors thus introduced are much too small to be considered.

Hence we may assume that the motion of the sun about the earth at rest is an ellipse having the earth for its focus, and its equation

$$u' = a'\{1 + e' \cos(\theta' - \zeta)\},$$

where $\zeta$ is the longitude of the sun's perigee and $e' = \frac{1}{60}$ is the eccentricity, and we are safe that no appreciable error will ensue in the determination of the moon's place.*

* That is, as far as the three bodies alone are concerned;—but, since the attractions of the planets may, and in fact do, disturb the elliptic orbit of the sun about $G$, the same cause will disturb the *assumed* orbit about $E$. A remarkable result of this disturbance is noticed in Appendix, Art. (108).

# CHAPTER III.

## RIGOROUS DIFFERENTIAL EQUATIONS OF THE MOON'S MOTION AND APPROXIMATE EXPRESSIONS OF THE FORCES.

19. The earth having been reduced to rest by the process described in Art. (9), let us take its centre $E$ for origin. Let $r$, $\theta$ be the polar coordinates of the projection $M'$ of the moon on the fixed plane of reference, $\theta$ being the longitude $\Upsilon EM'$ measured from a fixed line $E\Upsilon$ in that plane. Let $s$ be the tangent of the moon's latitude $MEM'$, so that $MM' = rs$.

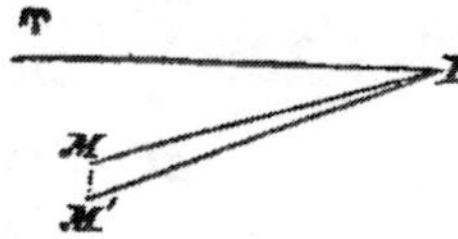


Next, let the accelerating forces which act upon the moon be resolved into these three:

$P$ parallel to the projected radius $M'E$ and *towards* the earth,

$T$ parallel to the fixed plane, perpendicular to $P$, and in the direction of $\theta$ *increasing*,

$S$ perpendicular to the fixed plane and *towards* it.

The equations of motion will be

$$\frac{1}{r}\frac{d}{dt}\left(r^2\frac{d\theta}{dt}\right) = T \text{ .................... (i)},$$

$$\frac{d^2r}{dt^2} - r\left(\frac{d\theta}{dt}\right)^2 = -P \text{ ..................(ii)},$$

$$\frac{d^2(rs)}{dt^2} = -S \text{ ................(iii)}.$$

20. These three equations for determining the moon's motion take the time $t$ for independent variable; but it will

be more convenient in the following process to consider the longitude as such, and our next step will be to change the independent variable from $t$ to $\theta$.

From (i) we get

$$r^2 \frac{d\theta}{dt} \frac{d}{dt}\left(r^2 \frac{d\theta}{dt}\right) = Tr^3 \frac{d\theta}{dt};$$

therefore
$$\left(r^2 \frac{d\theta}{dt}\right)^2 = h^2 + 2\int Tr^3 d\theta,$$

$$= H^2 \text{ suppose, whence } H\frac{dH}{d\theta} = Tr^3.$$

Therefore
$$\frac{d\theta}{dt} = \frac{H}{r^2} = Hu^2, \text{ if } u = \frac{1}{r},$$

$$\frac{dt}{d\theta} = \frac{1}{Hu^2} = \frac{1}{hu^2}\left(1 + 2\int \frac{T}{h^2u^3} d\theta\right)^{-\frac{1}{2}},$$

an important *equation connecting the time $t$ with the longitude $\theta$.*

Again,
$$\frac{dr}{dt} = \frac{dr}{d\theta} \cdot \frac{d\theta}{dt} = -H\frac{du}{d\theta},$$

$$\frac{d^2r}{dt^2} = \frac{d}{dt}\left(-H\frac{du}{d\theta}\right) = -Hu^2 \frac{d}{d\theta}\left(H\frac{du}{d\theta}\right).$$

Substitute in (ii),

therefore
$$Hu^2 \frac{d}{d\theta}\left(H\frac{du}{d\theta}\right) + H^2u^3 = P,$$

or
$$H^2u^2\left(\frac{d^2u}{d\theta^2} + u\right) + u^2\frac{du}{d\theta} H\frac{dH}{d\theta} = P \text{.........} (A);$$

and substituting for $H^2$ and for $H\frac{dH}{d\theta}$ their values $h^2 + 2\int \frac{T}{u^3} d\theta$ and $\frac{T}{u^3}$, then dividing by $h^2u^2$, and transposing,

$$\frac{d^2u}{d\theta^2} + u = \frac{P}{h^2u^2} - \frac{T}{h^2u^3}\frac{du}{d\theta} - 2\left(\frac{d^2u}{d\theta^2} + u\right)\int \frac{T}{h^2u^3} d\theta;$$

this is called the *differential equation of the moon's radius vector.*

Lastly,

$$-S=\frac{d^2\left(\frac{s}{u}\right)}{dt^2}=\frac{d}{dt}\left\{\frac{u\frac{ds}{d\theta}-s\frac{du}{d\theta}}{u^2}\cdot\frac{d\theta}{dt}\right\}=\frac{d}{dt}\left\{H\left(u\frac{ds}{d\theta}-s\frac{du}{d\theta}\right)\right\}$$

$$=H^2u^2\left(u\frac{d^2s}{d\theta^2}-s\frac{d^2u}{d\theta^2}\right)+H\frac{dH}{d\theta}u^2\left(u\frac{ds}{d\theta}-s\frac{du}{d\theta}\right);$$

from $(A)$, $$Ps=H^2u^2\left(s\frac{d^2u}{d\theta^2}+us\right)+H\frac{dH}{d\theta}u^2s\frac{du}{d\theta};$$

therefore $$Ps-S=H^2u^3\left(\frac{d^2s}{d\theta^2}+s\right)+H\frac{dH}{d\theta}u^3\frac{ds}{d\theta};$$

therefore

$$\frac{d^2s}{d\theta^2}+s=\frac{Ps-S}{h^2u^3}-\frac{T}{h^2u^3}\frac{ds}{d\theta}-2\left(\frac{d^2s}{d\theta^2}+s\right)\int\frac{T}{h^2u^3}d\theta;$$

this is the *differential equation of the moon's latitude.*

21. If these three equations could be integrated under these general forms, the problem of the moon's motion would be completely solved; for as only four variables $u$, $\theta$, $s$, and $t$ are involved (the accelerating forces $P$, $T$, and $S$ being functions of these), the values of three of them, as $u$, $\theta$, $s$, could be obtained in terms of the fourth $t$; that is, the radius vector, longitude, and latitude would be known corresponding to a given time.

But the integration has never yet been effected, except for particular values of $P$, $T$, and $S$; and the method which we are in consequence forced to adopt is that of successive approximation, by which the values of $u$, $\theta$, and $s$ are obtained in series in which the terms proceed according to ascending powers of certain small fractions. Some one fraction being chosen as a standard with which all other are compared, the order of the approximation is esteemed by the highest power of the small fractions retained.

It is usual to consider $\frac{1}{20}$ as a small fraction of the first order,
consequently $\frac{1}{20}$ of $\frac{1}{20} = \frac{1}{400}$ is ...................... second......
$\frac{1}{8000}$ is ...................... third ......
and so on, other fractions being considered as of the 1st, 2nd, &c. orders, according as they more nearly coincide with $\frac{1}{20}$, $\frac{1}{400}$, &c.

22. It is necessary therefore, before we can approximate at all, that we should have a previous knowledge (a rough one is sufficient) of the values of some of the quantities involved in our investigations; and for this knowledge we must have recourse to observation.

We shall therefore assume as data the following results of observation:

(1) The moon moves in longitude about thirteen times as fast as the sun. The ratio of the *mean* motions in longitude represented by $m$ is therefore about $\frac{1}{13}$, and may be considered as of the 1st order.*

(2) The sun's distance from the earth is about 400 times as great as the moon's distance.

Hence the ratio of the mean distances $= \frac{1}{400}$ is of the second order.†

(3) The eccentricity $e'$ of the elliptic orbit which the sun approximately describes about the earth is about $\frac{1}{60}$, and this, approaching nearer in value to $\frac{1}{20}$ than to $\frac{1}{400}$, will be considered as of the 1st order.

(4) During one revolution, the moon moves pretty accurately in a plane inclined to the plane of the ecliptic at

---

* This approximate value of $m$ is easily obtained;—the moon is found to perform the tour of the heavens, returning to the same position among the fixed stars, in about $27\frac{1}{3}$ days; the sun takes $365\frac{1}{4}$ days to accomplish the same journey.

† The distances of the luminaries may be calculated from their horizontal parallaxes, found by observations made at remote geographical stations. (See the Author's *Astronomy*, chap. XVI.).

an angle whose tangent is about $\frac{1}{12}$, and therefore of the 1st order.*

(5) Its orbit in this plane is very nearly an ellipse having the centre of the earth in its focus, and whose eccentricity is about equal to our standard of small fractions of the 1st order, viz. $\frac{1}{20}$; and this will also be very nearly true of the projection of the orbit on the plane of the ecliptic.†

*To calculate the values of the forces.*

23. We are now in possession of the data requisite for beginning our approximations, and we shall proceed to the determination of the values of $P$, $T$, and $S$ in terms of the coordinates of the positions of the sun and moon.

Let $S$, $E$, $M$ be the centres of the sun, earth, and moon,

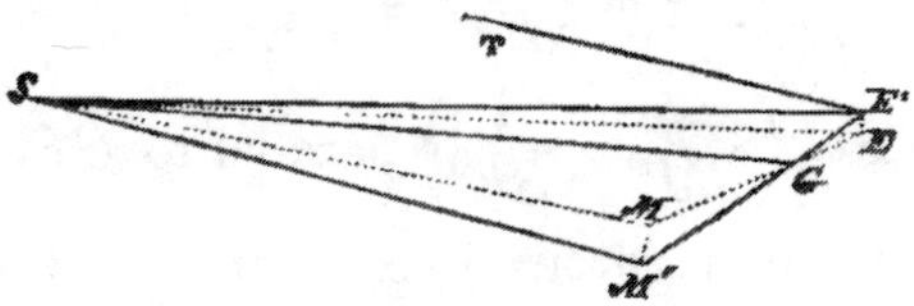


$m'$, $E$, $M$ their masses,

$E'$, $M'$, the projections on the plane of the ecliptic,

$G$ the centre of gravity of $E$ and $M$.

---

* That the moon's orbit during one revolution is very nearly a plane inclined as we have stated, will be found by noting her position day after day among the fixed stars; and the sun's path having previously been ascertained in a similar way, the rules of Spherical Trigonometry will easily enable us to verify these statements. (*Astronomy*, p. 92).

† The elliptic nature and the value of the eccentricity of the moon's orbit may be found by daily observation of her parallax, whence her distance from the earth's centre may be determined: corresponding observations of her place in the heavens being taken, and corrected for parallax to reduce them to the earth's centre, will determine her angular motion. Lines proportional to the distances being then drawn from a point in the proper directions, the extremities mark out the form of the moon's orbit.

A similar method applied to observations of the diameter of the sun will determine the eccentricity of its orbit. (*Astronomy*, p. 138).

The forces we have to take into account are, the forces which act directly on $M$, and forces equal and opposite to those which act on $E$;—these last being applied to the whole system so that $E$ may be a fixed point.

| *Forces on M.* | *Forces on E.* |
| --- | --- |
| Attraction of $E = \frac{E}{ME^2}$ ......in $ME$ | Attraction of $M = \frac{M}{ME^2}$......in $EM$ |
| attraction of $S = \frac{m'}{SM^2}$ in $MS$ equivalent to $\begin{cases} \frac{m'.MG}{SM^3} \text{ .....in } ME \\ \frac{m'.SG}{SM^3} \text{ par}^l\text{. to } GS \end{cases}$ | attraction of $S = \frac{m'}{SE^2}$ in $ES$ equivalent to $\begin{cases} \frac{m'.EG}{SE^3} \text{ .....in } EM \\ \frac{m'.SG}{SE^3} \text{ par}^l\text{. to } GS \end{cases}$ |

Therefore, the whole attraction upon $M$, when $E$ is brought to rest, is

$$\frac{E+M}{ME^2} + m'\left(\frac{MG}{SM^3} + \frac{EG}{SE^3}\right) \text{ in } ME,$$

and $$m'SG\left(\frac{1}{SM^3} - \frac{1}{SE^3}\right) \text{ parallel to } GS.$$

These are the rigorous expressions of the accelerating forces on $M$, and they can be expressed in terms of the masses and coordinates of the bodies.

*Approximate values of $P$, $T$, $S$.*

24. Our investigations will be carried only to the second order; it will be sufficient therefore if, in the preceding, we neglect small quantities of the fourth and higher orders.

Let $\mu = E + M$,

$MGM' = \tan^{-1}s =$ moon's latitude,

$E'\Upsilon$ the direction of the first point of Aries,

$SG = r' = \frac{1}{u'}$; $\angle \Upsilon E'S = \theta' =$ longitude of sun,

$M'E' = r = \frac{1}{u}$; $\angle \Upsilon E'M' = \theta =$ longitude of moon,

$\therefore$ $SE'M' = \theta - \theta' =$ difference of longitude of sun and moon.

Now, $SM^2 = SM'^2 + MM'^2$

$$= SG^2 + GM'^2 - 2SG.GM' \cos SGM' + MM'^2$$

$$= r'^2\left(1 - 2\frac{GM'}{r'}\cos SGM' + \frac{GM^2}{r'^2}\right);$$

therefore $$\frac{1}{SM^3} = \frac{1}{r'^3}\left\{1 + \frac{3GM'}{r'}\cos(\theta - \theta')\right\};$$

for $\theta - \theta'$ or $SE'M'$ differs from $SGM'$ by less than $7'$, Art. (17), and $\left(\frac{GM}{r'}\right)^2$ is neglected, being of the fourth order.

Similarly, $\frac{1}{SE'^3} = \frac{1}{r'^3}\left\{1 - \frac{3GE'}{r'}\cos(\theta - \theta')\right\}$; therefore the accelerating forces on the moon are approximately

$$\frac{\mu}{ME^2} + \frac{m'}{r'^3}(MG + GE) \text{ ......... in direction } ME,$$

and $\frac{3m'}{r'^3}(GM' + GE')\cos(\theta - \theta')$ ......... parallel to $GS$;

whence $$P = \left(\frac{\mu}{ME^2} + \frac{m'}{r'^3}ME\right)\cos MGM' - \frac{3m'}{r'^3}M'E'\cos^2(\theta - \theta')$$

$$= \frac{\mu}{r^2(1+s^2)^{\frac{3}{2}}} + \frac{m'r}{r'^3} - \frac{3m'r}{2r'^3}\{1 + \cos 2(\theta - \theta')\}$$

$$= \mu u^2(1 - \tfrac{3}{2}s^2) - \frac{m'u'^3}{u}\{\tfrac{1}{2} + \tfrac{3}{2}\cos 2(\theta - \theta')\},$$

$$T = -\frac{3m'}{r'^3}M'E'\cos(\theta - \theta')\sin(\theta - \theta')$$

$$= -\tfrac{3}{2}\frac{m'u'^3}{u}\sin 2(\theta - \theta'),$$

$$S = \left(\frac{\mu}{ME^2} + \frac{m'}{r'^3}ME\right)\sin MGM'$$

$$= \left\{\frac{\mu}{r^2(1+s^2)} + \frac{m'r\sqrt{(1+s^2)}}{r'^3}\right\}\frac{s}{\sqrt{(1+s^2)}}$$

$$= \mu u^2(s - \tfrac{3}{2}s^3) + \frac{m'u'^3 s}{u};$$

$$Ps - S = -\frac{m'u'^3 s}{u}\{\tfrac{3}{2} + \tfrac{3}{2}\cos 2(\theta - \theta')\}.$$

The differential equations in Art. (20), when these values of the forces are substituted in them, would contain the variables $u'$ and $\theta'$; but $u'$ is given in terms of $\theta'$ by the equation of Art. (18) $u' = a'\{1 + e' \cos(\theta' - \zeta)\}$, and we shall find means to establish a connexion between $\theta$ and $\theta'$, which will enable us to eliminate the latter.

25. Before proceeding further, it will be important to consider the order of the *disturbing* effect of the sun's action, compared with the direct action of the earth. If we examine the values of $P$, $T$, and $S$, it will be found that the most important of the terms containing $m'$, which are clearly the disturbing forces since they depend upon the sun, are involved in the form $\frac{m'r}{r'^3}$, while those independent of the sun's action enter in the form $\frac{\mu}{r^2}$.

We must therefore find the order of $\frac{m'r}{r'^3}$ compared with $\frac{\mu}{r^2}$,

$$\text{or of } \frac{m'}{r'^3} \;\ldots\ldots\ldots\ldots\ldots\ldots\; \frac{\mu}{r^3}.$$

Now the orbits being nearly circular, and $m$ the ratio of the mean motions, Art. (22), we have

$$m = \frac{\text{mean motion of sun}}{\text{mean motion of moon}} = \frac{\text{periodic time of moon}}{\text{periodic time of sun}}$$

$$= \frac{2\pi r^{\frac{3}{2}}}{\mu^{\frac{1}{2}}} \div \frac{2\pi r'^{\frac{3}{2}}}{m'^{\frac{1}{2}}};$$

therefore $$\frac{m'}{r'^3} : \frac{\mu}{r^3} :: m^2 : 1,$$

or the disturbing force of the sun is of the second order when compared with the direct action of the earth.

# CHAPTER IV.

## INTEGRATION OF THE DIFFERENTIAL EQUATIONS.

### SECTION I.

*General process described.*

26. If in the differential equations we substitute the values of $P$, $T$, and $S$ just obtained, we are led to the following forms:

$$\frac{d^2u}{d\theta^2} + u = F_1(u, \theta, s),$$

$$\frac{d^2s}{d\theta^2} + s = F_2(u, \theta, s),$$

$$\frac{dt}{d\theta} = F_3(u, \theta, s),$$

which cannot be integrated; but it is found possible to obtain a solution which will be a first approximation, if we retain only the important terms on the right-hand side of each equation. This first approximation being substituted in the equations, the terms next in order of importance may be retained and a new solution becomes possible, which in its turn is the basis of a further approximation, and so on.

First, neglect the disturbing force of the sun which is of the second order, and also the moon's latitude, which, as will be seen by referring to the expressions for the forces (Art. 24), will either enter to the second power or else in combination with the disturbing force.

When this is done the equations become integrable, and values of $u$ and $s$ may be obtained in terms of $\theta$ correct to the first order; this value of $u$ will then enable us to get the connexion between $\theta$ and $t$ to the same order.

[Let us, however, bear in mind that the equations thus integrated are not the differential equations of the moon's motions, but only approximate forms of them; and it is, therefore, possible that the results obtained may be approximations to the true solutions only for a limited range of values.

Whether they are so or not, will appear by comparing them with what we already know of the motion from observation; and this previous knowledge, in the event of their not being approximations, will probably suggest such modifications of them as will render them so.]

To proceed to the next approximation:—We must simplify the right-hand members of our general equations by employing the values of $u$ and $s$ just obtained, retaining now terms of the second order which we had previously neglected.

The integration can always be performed when the right-hand members of the equations are circular functions (sines or cosines) of $\theta$ or of its multiples; and as our first approximation will give us the values of $u$ and $s$ in that form, the new expressions for the forces will also be of the same character, and the equations be again integrable.

Thus we shall obtain new values of $u$, $s$, $t$ correct to the second order. These values, introduced in the same manner in the second members and terms of the next higher order retained, will lead to a third approximation, and so on, to any order; except that if we wish to carry it on beyond the third, the approximate values of the forces, given in Art. (24), would no longer be sufficiently exact, and we must obtain more correct values from the expressions of Art. (23).

27. We must here stop to notice a peculiarity in these equations, when solved by this process. We have said that

to obtain the values to any order, all terms up to that order must be retained in the second members; but it may happen that a term of an order beyond that to which we are working would, if retained, be so altered by the integration as to come within the proposed order.

Such terms must therefore not be rejected, and we shall proceed to examine by what means they may be recognised.

28. Suppose then that after an approximation to a certain order, the substitutions for the next steps have brought the equation in $u$ to the form

$$\frac{d^2u}{d\theta^2} + u = \ldots\ldots + G\cos(p\theta + H) + \ldots\ldots,$$

where the coefficient $G$ is one order beyond that which we intend to retain. The solution of this equation will be of the form

$$u = \ldots\ldots + G' \cos(p\theta + H) + \ldots\ldots$$

$G'$ being a constant which may be determined by substituting this value of $u$ in the differential equation,

whence $$G' = \frac{G}{1-p^2};$$

from which we learn that if $p$ differs very little from 1, $G'$ will be at least one order *lower* than $G$, and will come within our proposed approximation; and consequently the term $G\cos(p\theta + H)$ must be retained in the differential equation.

The equation of the moon's latitude being of the same form as that of the radius vector, the same remarks apply to it.

29. If $p$ is small, $G' = G$ very nearly, and the term $G\cos(p\theta + H)$ will not rise in importance in the value of $u$.

We should therefore be led to reject it in the differential equation, except for the following reason:—

In finding the connexion between the longitude and the time (one of the principal objects of the Theory), we must use the equation

$$\frac{dt}{d\theta} = \frac{1}{hu^2}\left(1 + 2\int\frac{T}{h^2u^3}d\theta\right)^{-\frac{1}{2}}.$$

Now, substituting for $u$ its value in terms of $\theta$, we shall find that such a term as $G' \cos(p\theta + H)$ in $u$ gives rise to a term $G'' \cos(p\theta + H)$ in this equation, where $G'$ and $G'$ are of the same order, then

$$\frac{dt}{d\theta} = \ldots\ldots G'' \cos(p\theta + H),$$

whence

$$t = \ldots\ldots\frac{G''}{p}\sin(p\theta + H);$$

therefore, when $p$ is a small quantity of the first order, $\frac{G''}{p}$ will be one order lower than $G''$, and the term will have risen in importance by the integration.

But yet further, if such terms occur in $\frac{T}{h^2u^3}$, they will be twice increased in value; for they increase once in forming $\int\frac{T}{h^2u^3}d\theta$, and once again, as above, in finding $t$.

30. We have, therefore, the following rule:—

*In approximating to any given order to the values of u and s, we must, in their differential equations, retain periodical terms* ONE ORDER *beyond the proposed one, when the coefficient of θ in their argument* is near unity.*

---

* The angle of a periodical term is its only variable part and is called the *argument*.

*In approximating to the value of t, we must, in the same equations, also retain terms* ONE ORDER *beyond the proposed one when the coefficient of the argument is near zero; and when such terms occur in* $\frac{T}{h^2u^3}$ *they must be retained* TWO ORDERS *beyond the proposed approximation.**

31. We shall here, for convenience of reference, bring together the equations and the approximate expressions for the forces:—

$$\frac{d^2u}{d\theta^2}+u=\frac{P}{h^2u^2}-\frac{T}{h^2u^3}\cdot\frac{du}{d\theta}-2\left(\frac{d^2u}{d\theta^2}+u\right)\int\frac{T}{h^2u^3}\,d\theta \;\ldots\; (A),$$

$$\frac{d^2s}{d\theta^2}+s=\frac{Ps-S}{h^2u^3}-\frac{T}{h^2u^3}\cdot\frac{ds}{d\theta}-2\left(\frac{d^2s}{d\theta^2}+s\right)\int\frac{T}{h^2u^3}\,d\theta \;\ldots\; (B),$$

$$\frac{dt}{d\theta}=\frac{1}{hu^2}\left(1+2\int\frac{T}{h^2u^3}\,d\theta\right)^{-\frac{1}{2}} \;\ldots\ldots\ldots\ldots\; (C),$$

$$\frac{P}{h^2u^2}=\frac{\mu}{h^2}\left(1-\tfrac{3}{2}s^2\right)-\frac{m'u'^3}{2h^2u^3}\{1+3\cos 2(\theta-\theta')\} \;\ldots\ldots\; (D),$$

$$\frac{T}{h^2u^3}=-\frac{3}{2}\frac{m'u'^3}{h^2u^4}\sin 2(\theta-\theta') \;\ldots\ldots\ldots\ldots\; (E),$$

$$\frac{Ps-S}{h^2u^3}=-\frac{3}{2}\frac{m'u'^3s}{h^2u^4}\{1+\cos 2(\theta-\theta')\} \;\ldots\ldots\; (F).$$

---

* Instead of the forces which really act on the moon, we originally substituted three equivalent ones, $P$, $T$, $S$; these again are, by the preceding expressions, replaced by a set of others. For, we may conceive each of the terms in $\frac{P}{h^2u^3}$, &c., to correspond to a force,—a component of $P$, $T$, or $S$; each force having the same argument as the term to which it corresponds, and therefore going through its cycle of values in the same time. Now, by Art. (29), when the coefficient of $\theta$ in the argument is near unity, the term becomes important in the radius vector, and when near zero, in the longitude: hence, a force whose period is nearly the same as that of the moon, produces important effects in the radius vector; and a force whose period is very long will be important in its effect on the longitude. See *Airy's* Tracts, *Planetary Theory*, p. 78.

## Section II.

*To solve the Equations to the first order.*

32. Neglect all terms depending on the disturbing force, *i.e.* those which contain $m'$; such terms being of the second order. Art. (25).

The latitude $s$ of the moon can never exceed the inclination of the orbit to the ecliptic; but this inclination is of the first order, Art. (22), therefore $s$ is at least of the first order and $s^2$ may be neglected.

Whence $$\frac{P}{u^2} = \mu; \quad \frac{T}{u^3} = 0; \quad \frac{Ps - S}{u^3} = 0,$$

and the differential equations become

$$\frac{d^2u}{d\theta^2} + u = \frac{\mu}{h^2},$$

$$\frac{d^2s}{d\theta^2} + s = 0;$$

whence $u = \frac{\mu}{h^2}\{1 + e\cos(\theta - \alpha)\}$; or, writing $a$ for $\frac{\mu}{h^2}$,

$$u = a\,\{1 + e\cos(\theta - \alpha)\} \quad \ldots\ldots\ldots\ldots\ldots\ldots \quad (U_1),$$

and $$s = k\sin(\theta - \gamma) \quad \ldots\ldots\ldots\ldots\ldots\ldots\ldots \quad (S_1),$$

$e$, $\alpha$, $k$, $\gamma$ being the four constants introduced by integration.

33. These results are in perfect agreement with what *rough* observations had already taught us concerning the moon's motion Art. (22); for $u = a\{1 + e\cos(\theta - \alpha)\}$ represents motion in an ellipse about the earth as focus.

And $s = k\sin(\theta - \gamma)$ indicates motion in a plane inclined to the ecliptic at an angle $\tan^{-1}k$.

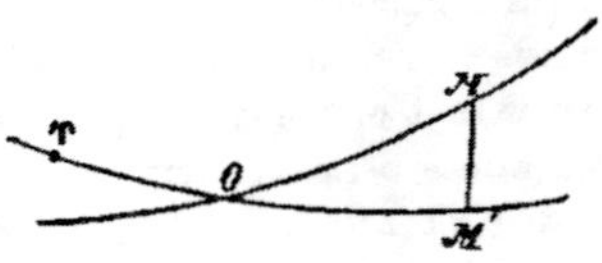

For, if $\Upsilon OM'$ be the ecliptic, $M$ the moon's place, $MM'$ an arc perpendicular to the ecliptic, then

$\Upsilon M' = \theta$; and if $\Upsilon O$ be taken equal to $\gamma$, and $OM$ joined by an arc of great circle, we have

$$\tan MM' = \tan MOM' \sin OM';$$

or
$$s = \tan MOM' \sin(\theta - \gamma),$$

which, compared with the equation above, shews that

$$MOM' = \tan^{-1} k.$$

Therefore the moon moves in a plane passing through a fixed point $O$ and making a constant angle with the ecliptic.

The values of $\alpha$ and $\gamma$ introduced in the above solutions are respectively the longtitude of the apse and of the node.

34. What the equations cannot teach us, however, is the magnitude of the quantities $e$ and $k$. For this we must have recourse to observation, and by referring to Art. (22), we see that $e$ is about $\frac{1}{20}$ and $k$ about $\frac{1}{12}$, that is, both quantities are of the first order. Their exact values cannot yet be obtained: the means of doing so from multiplied observations will be indicated further on.

35. Lastly, to find the connexion between $t$ and $\theta$, the equation ($C$) becomes, $T = 0$,

$$\frac{dt}{d\theta} = \frac{1}{hu^2} = \frac{1}{ha^2} \frac{1}{\{1 + e\cos(\theta - \alpha)\}^2}.$$

Now this is the very same equation that we had found connecting $t$ and $\theta$ in the problem of *two bodies*, Art. (12), as we ought to expect, since we have neglected the sun's action. Therefore, if $p$ be the moon's mean angular velocity, we should, following the same process as in the article referred to, arrive at the result

$$\theta = pt + \varepsilon + 2e\sin(pt + \varepsilon - \alpha) + \tfrac{5}{4}e^2 \sin 2(pt + \varepsilon - \alpha) + \ldots\ldots,$$

which is correct only to the first order, since we have

rejected some terms of the second order by neglecting the disturbing force.

36. The arbitrary constant $\varepsilon$, introduced in the process of integration, can be got rid of by a proper assumption: this assumption is, that the time $t$ is reckoned from the instant when the *mean* value of $\theta$ is zero.*

For, since the *mean* value of $\theta$, found by rejecting the periodical terms, is $pt + \varepsilon$; if, when this vanishes, $t = 0$, we must have $\varepsilon = 0$; therefore

$$\theta = pt + 2e \sin(pt - \alpha) \quad\ldots\ldots\ldots\ldots\ldots\ldots \Theta_1,$$

correct to the first order.†

37. We have now obtained three results, $U_1$, $S_1$, $\Theta_1$, as solutions to the first order of our differential equations, and we must employ them to obtain the next approximate solutions; but, before they can be so employed, they will require to be slightly modified—in such a manner, however, as not to interfere with their degree of approximation.

The necessity for such a modification will appear from the following considerations:

---

* When a function of a variable contains periodical terms which go through all their changes positive and negative as the variable increases continuously, the *mean value* of the function is the part which is independent of the periodical terms.

† We shall also employ this method of correcting the integral in our next approximation to the value of $\theta$ in terms of $t$; and if we purposed to carry our approximations to a higher order than the second, we should still adopt the same value, that is, zero, for the arbitrary constant introduced by the integration. To shew the advantage of thus correcting with respect to *mean* values: suppose we reckoned the time from some *definite* value of $\theta$, for instance when $\theta = 0$; then, in the first approximation,

$$0 = \epsilon + 2e \sin(\epsilon - \alpha)$$

is the equation for determining the constant $\epsilon$, and in the second approximation, $\epsilon$ would be found from

$$0 = \epsilon + 2e \sin(\epsilon - \alpha) + \tfrac{5}{4}e^2 \sin 2(\epsilon - \alpha) + \ldots\ldots,$$

giving different values of $\epsilon$ at each successive approximation.

Suppose we proceed with the values already obtained; we have, by Art. (24),

$$\frac{P}{h^2u^2} = \frac{\mu}{h^2}(1 - \tfrac{3}{2}s^2) - \frac{m'u'^3}{2h^2u^3} - \&c.......$$

$$= a(1 - \tfrac{3}{2}s^2) - \frac{m'u'^3}{2h^2a^3}\{1 + e\cos(\theta - \alpha)\}^{-3}$$

$$= a + ............ + A\cos(\theta - \alpha) + ...... ;$$

and this being substituted in the differential equation $(A)$ of Art (31), gives

$$\frac{d^2u}{d\theta^2} + u = a + ........... + A\cos(\theta - \alpha) + ......,$$

the solution of which is

$$u = a\{1 + e\cos(\theta - \alpha)\} + ...... + \tfrac{1}{2}A\theta\sin(\theta - \alpha).$$

Our first approximate value $u = a\{1 + e\cos(\theta - \alpha)\}$ is here corrected by a term which, on account of the factor $\theta$, admits of indefinite increase, and thus becomes ultimately more important than that with which we started. Such a correction is inadmissible, for our first value is confirmed by observation (22) to be very nearly the true one. The moon's distance, as determined by her parallax, is never much less than 60 times the earth's radius; whereas this new value of $u$, when $\theta$ is very great, would make the distance indefinitely small. On the same principle, we see that any solution, which comprises a term of the form $A\theta\sin(\theta - \alpha)$, cannot be an approximate solution except for a small range of values of $\theta$.

Snch terms, 'if they really had an existence in our system, mnst end in its 'destruction, or at least in the total subversion of its present state; but when 'they do occnr, they have their origin, not in the nature of the differential 'equations, hut in the imperfection of our analysis, and in the inadequate repre-'sentation of the perturbations, and are to be got rid of, or rather included in 'more general expressions of a periodical nature, by a more refined investigation 'than that which led ns to them. The nature of this difficulty will be easily 'understood from the following reasoning. Suppose that a term, such as '$a\sin(A\theta + B)$, should exist in the value of $u$, in which $A$ being extremely 'minute, the period of the inequality denoted by it would be of great length; 'then, whatever might be the value of the coefficient $a$, the inequality would 'still be always confined within certain limits, and after many ages would return 'to its former state.

'Suppose now that our peculiar mode of arriving at the value of $u$ led us 'to this term, not in its real analytical form $a\sin(A\theta+B)$, but by the way of 'its development in powers of $\theta$, $\alpha+\beta\theta+\lambda\theta^2+$&c.; and that, not at once, but 'piecemeal, as it were; a first approximation giving us only the term $\alpha$, a 'second adding the term $\beta\theta$, and so on. If we stopped here, it is obvious that 'we should mistake the nature of this inequality, and that a really periodical 'function, from the effect of an imperfect approximation, would appear under 'the form of one not periodical......... These terms in the value of $u$, when they 'occur, are not superfluous; they are essential to its expression, but they lead 'us to erroneous conclusions as to the stability of our system and the general 'laws of its perturbations, unless we keep in view that *they are only parts of 'series*; the principal parts, it is true, when we confine ourselves to intervals 'of moderate length, but which cease to be so after the lapse of very long times, 'the rest of the series acquiring ultimately the preponderance, and compensating 'the want of periodicity of its first terms.—SIR JOHN HERSCHEL, *Encyclopædia Metropolitana*—PHYSICAL ASTRONOMY, p. 679.

38. To extritcate ourselves from this difficulty, and to alter the solution so that none but periodical terms may be introduced, let us again observe that the equation $\frac{d^2u}{d\theta^2}+u=\frac{\mu}{h^2}=a$, which gave the solution $U_1$ and thus led to the difficulty, is only an approximate form of the first order of the exact equation. Any value of $u$, therefore, which satisfies the approximate equation $\frac{d^2u}{d\theta^2}+u=a$ to the first order, and which evades the difficulty mentioned above, may be taken as a solution to the same order of the exact equation.

These conditions will be satisfied if we assume

$$u=a\{1+e\cos(c\theta-\alpha)\};$$

then

$$\frac{d^2u}{d\theta^2}+u=a+ae(1-c^2)\cos(c\theta-\alpha)$$

$$=a \text{ to the first order,}$$

provided $1-c^2$ be of the first order at least.

39. The introduction of the factor $c$ in the argument is an artifice due to Clairaut; and its effect, as we shall see in Art. (66) is equivalent to supposing that the apse of the moon's orbit is not fixed. Now, although the observations recorded in Art. (22) did not suggest this, we must bear in mind that they were extremely rough, and carried on only for a short interval; but when they are made with a little more accuracy, and extended over several revolutions of the moon, it is soon found that both her apse and the plane of her orbit are in constant motion.

Clairaut was fully aware of these motions, and there is no doubt that he was led to the above form of the value of $u$ by that consideration, and by his acquaintance with the results of Newton's ninth section, which, when translated into analytical language, lead at once to the same form.*

We might, therefore, taking for granted the results of observation, have commenced our approximation at this step, and have at once written down $u = a\{1 + e\cos(c\theta - \alpha)\}$; but we should, in so doing, have merely postponed the difficulty to the next step, since there again, as we shall find, the differential equation is of the form

$$\frac{d^2u}{d\theta^2} + u = \text{a function of } \theta,$$

---

* Newton has there shewn, that if the angular velocity of the orbit be to that of the body as $G - F$ to $G$, the additional centripetal force is $\frac{G^2 - F^2}{G^2} h^2u^3$, the original force being $\mu u^2$. Therefore

$$\frac{d^2u}{d\theta^2} + u = \frac{\mu}{h^2} + \frac{G^2 - F^2}{G^2} u,$$

$$\frac{d^2u}{d\theta^2} + \frac{F^2}{G^2} u = \frac{\mu}{h^2} = \frac{F^2}{G^2}\frac{\mu}{h_{,}^2} = \frac{F^2}{G^2} a,$$

$$u = a\left\{1 + e\cos\left(\frac{F}{G}\theta - \alpha\right)\right\},$$

where $\frac{F}{G}$ is the same as our $c$.

the correct integral of which would be

$$u = A\cos(\theta - B) + \dots\dots\dots\dots,$$

and this would at the next operation bring in a term with $\theta$ for a coefficient, which we *now* know must not be. We shall, therefore, hereafter omit such terms as $A\cos(\theta - B)$ altogether, and merely write

$$u = a\{1 + e\cos(c\theta - \alpha)\} + \dots\dots\dots\dots.$$

40. The value of $c$ will be found more and more correctly at each successive approximation, by always assuming these terms as the first two terms of the value of $u$, then substituting in the differential equation and equating coefficients. It will thus be found that $ae(1 - c^2)$ must equal the coefficient of $\cos(c\theta - \alpha)$ in the differential equation; and this will enable us to determine $c$ to the same degree of approximation as that of the differential equation itself. See Arts. (52) and (103). So far, all that we know about $c$ is that it differs from unity at most by a quantity of the first order.

41. In carrying on the solution of $s$, the same difficulty arises as in $u$, and it will be found necessary to change it into

$$s = k\sin(g\theta - \gamma),$$

$g$ being a quantity which differs from unity at most by a quantity of the first order.

The introduction of $g$ is connected with the motion of the node in the same way as that of $c$ is with the motion of the apse (85); and the value of $g$ will be determined by a process similar to that explained above for $c$. Arts. (51) and (102).

42. The connexion between $\theta$ and $t$ will also be modified by this change in the value of $u$,

$$\frac{dt}{d\theta} = \frac{1}{ha^2}\,\frac{1}{\{1 + e\cos(c\theta - \alpha)\}^2},$$

or

$$\frac{d.ct}{d.c\theta} = \frac{1}{ha^2}\,\frac{1}{\{1 + e\cos(c\theta - \alpha)\}^2}.$$

Here $c\theta$ and $ct$ hold the places which $\theta$ and $t$ occupied in (35); therefore

$$c\theta = cpt + 2e \sin(cpt - \alpha),$$

or

$$\theta = pt + 2e \sin(cpt - \alpha),$$

to the first order, since $\frac{e}{c} = e$ to the first order.

43. Since the disturbing forces are to be taken into account in the next approximation, we shall have to use the value of $u'$ found in (18), which is

$$u' = a' \{1 + e' \cos(\theta' - \zeta)\}:$$

but this introduces $\theta'$; we must therefore further modify it by substituting for $\theta'$ its value in terms of $\theta$, and it will be found sufficient, for the purpose of the present work, to obtain the connexion between them to the first order, which may be done as follows:

Let $m$ be the ratio of the mean motions of the sun and moon,

$p'$, $p$ their mean angular velocities; $\therefore p' = mp$,

$p't + \beta, pt$ ...... mean longitudes at time $t$, $\beta$ being the sun's longitude when $t = 0$,

$\theta'$, $\theta$ ...... true longitudes at time $t$,

$\zeta$, $\alpha$ ...... longitude of perigees when $t = 0$;

therefore $\theta' - \zeta =$ sun's true anomaly,

and $p't + \beta - \zeta =$ ...... mean anomaly.

But, by Art. (13),

true anomaly = mean anomaly + $2e'$ sin (mean anomaly) + &c.;

$$\begin{aligned} \text{therefore} \quad \theta' &= p't + \beta + 2e' \sin(p't + \beta - \zeta) + \ldots\ldots \\ &= mpt + \beta + 2e' \sin(mpt + \beta - \zeta) + \ldots\ldots \\ &= m\theta + \beta + 2e' \sin(m\theta + \beta - \zeta) \end{aligned}$$

to the first order;

because $pt = \theta - 2e \sin(c\theta - \alpha)$ to the first order by (42).

Whence $u' = a' \{1 + e' \cos(m\theta + \beta - \zeta)\}$ to the first order.

44. The values of $\sin 2(\theta - \theta')$ and $\cos 2(\theta - \theta')$ can also be readily obtained to the same order:

$$\sin 2(\theta - \theta')$$
$$= \sin\{(2 - 2m)\theta - 2\beta - 4e' \sin(m\theta + \beta - \zeta)\}$$
$$= \sin\{(2 - 2m)\theta - 2\beta\} - 4e' \sin(m\theta + \beta - \zeta) \cos\{(2 - 2m)\theta - 2\beta\}$$
$$= \sin\{(2 - 2m)\theta - 2\beta\} - 2e' \sin\{(2 - m)\theta - \beta - \zeta\}$$
$$+ 2e' \sin\{(2 - 3m) - 3\beta + \zeta\}.$$

Similarly, $\cos 2(\theta - \theta') = \cos\{(2 - 2m)\theta - 2\beta\}$

$$- 2e' \cos\{(2 - m)\theta - \beta - \zeta\} + 2e' \cos\{(2 - 3m)\theta - 3\beta + \zeta\}.$$

The first term of each of these is all we shall require.

## SECTION III.

*To solve the equations to the Second Order.*

45. Let us recapitulate the results of the last approximation.

$$u = a\{1 + e\cos(c\theta - \alpha)\},$$
$$u' = a'\{1 + e'\cos(m\theta + \beta - \zeta)\},$$
$$s = k\sin(g\theta - \gamma),$$
$$\theta - \theta' = (1 - m)\theta - \beta - 2e' \sin(m\theta + \beta - \zeta).$$

These values must now be substituted in the expressions for the forces, retaining terms above the second order, when, according to the criterion of Art. (30), they promise to become of the second order after integrating.

The differential equations ($A$) and ($B$) will then assume the forms

$$\frac{d^2u}{d\theta^2} + u = F(\theta),$$

$$\frac{d^2s}{d\theta^2} + s = f(\theta),$$

and the integration of these will enable us to obtain $u$ and $s$ to the second order; after which, equation ($C$) will give the connexion between $\theta$ and $t$ to the same order.

46. The quantity $\frac{m'a'^3}{h^2a^3}$, which we shall meet with as a coefficient of the terms due to the disturbing force, can be replaced by $m^2a$, $m$ being the ratio of the mean motions of the sun and moon.

So long as we neglected the disturbing force, $h$ and $a$ had determinate values:—they belonged to the ellipse which formed our first imperfect solution, and would therefore be known from the circumstances of motion in that ellipse at any instant; $h$ being double the area described in a unit of time, and $a$ the reciprocal of the semi-latus rectum. It would consequently be impossible to assume any *arbitrary* connexion between them. But, when we proceed to a second approximation and introduce the disturbing force, there is no longer a determinate ellipse to which the $h$ and $a$ apply: the equation $\mu = h^2a$ of Art. (32) merely shews that $a$ and $h$ must refer to some one of the instantaneous ellipses which the moon could describe about the earth if the disturbance were to cease, and we are at liberty to select any one of these which will allow us to proceed with our approximation. The particular ellipse is determined by the above assumed relation $\frac{m'a'^3}{h^2a^3} = m^2a$, and the selection is suggested and justified by the following reasoning:

$$m = \frac{p'}{p} = \frac{\text{average period of moon about earth}}{\text{period of sun about earth}},$$

but, since the instantaneous ellipses are nearly circles, we have, as in Art. (25),

$$\frac{(\text{period of moon about earth})^2}{(\text{period of sun about earth})^2} = \frac{m'a'^3}{\mu a^3}, \text{ nearly},$$

therefore if $a$ be properly chosen,

$$m^2 = \frac{m'a'^3}{\mu a^3} = \frac{m'a'^3}{h^2 a^4}.$$

47. If we examine the equations of Art. (31) it will be seen that we cannot solve the differential equation in $u$ to the second order until we know the value of $s$ accurately to that order; for, suppose

$$s = k \sin(g\theta - \gamma) + V_2,$$

where $V_2$ contains the terms of the second order not yet found; and suppose that among the terms in $V_2$ there be one $L \sin(l\theta - \lambda)$, then this value of $s$ substituted in the expression $(D)$ gives

$$\frac{P}{h^2u^2} = \frac{\mu}{h^2}\{1 - \tfrac{3}{2}[k \sin(g\theta - \gamma) + L \sin(l\theta - \lambda)\ldots]^2\} - \ldots$$

$$= a\{1 - \tfrac{3}{2}k^2 \sin^2(g\theta - \gamma) - 3kL \sin(g\theta - \gamma) \sin(l\theta - \lambda)\ldots\},$$

and equation $(A)$ becomes, so far as this term is concerned,

$$\frac{1}{a}\left(\frac{d^2u}{d\theta^2} + u\right) = 1 + \ldots - \tfrac{3}{2}kL[\cos\{(g - l)\theta - (\gamma - \lambda)\}]$$

$$+ \tfrac{3}{2}kL[\cos\{(g + l)\theta - (\gamma + \lambda)\}]\ldots.$$

Now, if it should happen that $l$ is a small quantity, then the two coefficients $g - l$ and $g + l$ will be both near unity, and therefore these two terms, although of the third order in the differential equation, would become important in the value of $u$. Again, if $l$ is near 2, then $l-g$ will be near unity, and the same remark will apply.

Such terms do not really occur in the value of $s$, but there is none the less a necessity for verifying the fact.

If $l$ is near unity, then the coefficient $g-l$ becomes small; and although the corresponding term in $u$ is still of the third order, it becomes of the second order, and therefore important in $t$, Art. (30).

If $l$ is not near 0, 1 or 2, the term will not affect, to that order, the values of $u$ or of $t$.

48. Again, it is necessary to have the value of $u$ correct to the second order, before we can compute $\frac{T}{h^2u^3}$ to the fourth order, with a full assurance that we have secured all terms of that order in which the coefficient of the argument is nearly zero*;—such terms becoming, as we have seen Art. (30), important in the value of $t$.

For, assume that

$$u = a\{1 + e\cos(c\theta - \alpha) + W_2\},$$

then $$\frac{T}{h^2u^3} = -\frac{3m'a'^3\{1+e'\cos(m\theta+\beta-\zeta)\}^3}{2h^2a^4\{1+e\cos(c\theta-\alpha)+W_2\}^4}\sin\{(2-2m)\theta-2\beta\}$$

$$= -\tfrac{3}{2}m^2(1\ldots-4W_2)\sin\{(2-2m)\theta-2\beta\}.$$

Therefore if $W_2$ contain a term $R\cos(r\theta-\rho)$ of the second order where $r$ is near 2 (except $2-2m$) there will be a

* This was pointed out by Mr. Walton in the *Quarterly Journal of Mathematics*, Vol. IX., p. 227; and in this edition the text had been modified in accordance with his suggestion; but the method given by him for finding the value of $u$ is defective on account of its not mentioning the necessity for the previous determination of $s$.

corresponding term in $\frac{T}{h^2u^3}$ of the fourth order with a small coefficient of $\theta$, and therefore an important term in the value of $t$.

49. For the sake of simplifying the expressions we shall generally omit $\alpha$, $\beta$, and $\gamma$ in the arguments. If we remark that $c$ and $\alpha$ always appear together in the form $c\theta-\alpha$, it will be sufficient to write $c\theta_1$ instead of $c\theta-\alpha$; and, for the same reason, we may replace $m\theta+\beta$ and $g\theta-\gamma$ by $m\theta_1$ and $g\theta_1$. The suffix ($_1$) will remind us of the omission, and we may at any stage re-introduce the omitted symbols.

*To compute s to the second order.*

50. Since we must first determine the value of $s$, it becomes necessary to examine the differential equation ($B$) to see whether any difficulty of a similar kind to those we have been here considering may arise, or whether the elements on which the computation depends are already known with sufficient accuracy. We fortunately find that all the terms of the right-hand member of the equation are of the third order, and therefore that the values already obtained will ensure accuracy to the second order, and we may safely begin our work here.

51. To the first order we have

$$s = k \sin g\theta_1,$$

$$\frac{ds}{d\theta} = kg \cos g\theta_1 = k \cos g\theta_1 \begin{cases} \text{because } g=1 \\ \text{to first order,} \end{cases}$$

$$\frac{d^2s}{d\theta^2} + s = 0.$$

In the expression for $\frac{Ps - S}{h^2u^3}$, which is of the third order, it will be sufficient to put $a'$ for $u'$, and $a$ for $u$, then

$$\frac{Ps - S}{h^2u^3} = -\tfrac{3}{2}\,\frac{m'a'^3}{h^2a^4}\,k\sin g\theta_1\,\{1 + \cos(2 - 2m)\,\theta_1\}$$

$$= -\tfrac{3}{2}m^2k\sin g\theta_1 + \tfrac{3}{4}m^2k\sin(2 - 2m - g)\,\theta_1,$$

the other term of the third order being rejected because the coefficient of $\theta$ is not near unity.

The next term of equation ($B$) is $\frac{T}{h^2u^3}\frac{ds}{d\theta}$, and as $\frac{ds}{d\theta}$ is of the first order, it will be sufficient to have $\frac{T}{h^2u^3}$ correct to the second order,

$$\frac{T}{h^2u^3} = -\tfrac{3}{2}m^2\sin(2 - 2m)\,\theta_1,$$

$$\frac{T}{h^2u^3}\cdot\frac{ds}{d\theta} = -\tfrac{3}{2}m^2k\sin(2 - 2m)\,\theta_1\cos g\theta_1$$

$$= -\tfrac{3}{4}m^2k\sin(2 - 2m - g)\,\theta_1,$$

neglecting the other term of the third order because the coefficient of $\theta$ is not near unity.

The last term of equation ($B$) is $2\left(\frac{d^2s}{d\theta^2} + s\right)\int\frac{T}{h^2u^3}\,d\theta$ which $= 0$ to the third order, since $\int\frac{T}{h^2u^3}\,d\theta$ depending on the disturbing force is of the second order, and the other factor $= 0$ to the first order.

| The equation in $s$ becomes | To solve it, assume |
|---|---|
| $\frac{d^2s}{d\theta^2} + s =$ | $s =$ |
| $-\tfrac{3}{2}m^2k\sin g\theta_1$ | $k\sin g\theta_1$ |
| $+\tfrac{3}{4}m^2k\sin(2 - 2m - g)\,\theta_1.$ | $+ A\sin(2 - 2m - g)\,\theta_1.$ |

Substitute, and equate coefficients of like terms,

$$k(1-g^2) = -\tfrac{3}{2}m^2k;\ \therefore\ g = (1+\tfrac{3}{2}m^2)^{\frac{1}{2}} = 1 + \tfrac{3}{4}m^2,$$

$$A\{1-(2-2m-g)^2\} = \tfrac{3}{2}m^2k;\ \therefore\ A = \frac{\tfrac{3}{2}m^2k}{1-(1-2m-\tfrac{3}{4}m^2)^2} = \tfrac{3}{8}mk.$$

Therefore, accurately, to the second order,*

$$s = k\sin(g\theta-\gamma) + \tfrac{3}{8}mk\sin\{(2-2m-g)\,\theta - 2\beta + \gamma\}\ldots S_2.$$

The term of the second order, having a coefficient near unity, need not be taken into account in finding the value of $u$ (47); but it will have to be brought in subsequently to determine $t$.

*To determine u to the second order.*

52. We must compute the right-hand member of the equation ($A$), Art. (31), accurately to the second order, and include those terms of the third order which have a coefficient near unity.

To the first order we have

$$u = a(1 + e\cos c\theta_1),$$

$$\frac{du}{d\theta} = -aec\sin c\theta_1 = -ae\sin c\theta_1 \quad \begin{cases}\text{because } c = 1\\ \text{to first order,}\end{cases}$$

$$\frac{d^2u}{d\theta^2} + u = a.$$

From the expressions ($D$) and ($E$), Art. (31), we get

$$\frac{P}{h^2u^2} = a(1 - \tfrac{3}{2}k^2\sin^2 g\theta_1)$$
$$- \frac{m'a'^3\{1+e'\cos(m\theta_1-\zeta)\}^3}{2h^2a^3(1+e\cos c\theta_1)^3}\{1 + 3\cos(2-2m)\,\theta_1\},$$

* No complementary term $P\cos(\theta - Q)$ is added; for, though by the theory of differential equations, this would form a necessary part of the solution, we have seen Art. (39) that it cannot, in this shape, form a part of the correct value, but will be comprised in the terms whose argument is $g\theta-\gamma$.

$$\frac{1}{a}\frac{P}{h^2u^2} = 1 - \tfrac{3}{4}k^2 + \tfrac{3}{4}k^2 \cos 2g\theta_1$$

$$- \tfrac{1}{2}m^2 \{1 + 3e' \cos(m\theta_1 - \zeta) - 3e \cos c\theta_1\} \{1 + 3 \cos(2 - 2m)\,\theta_1\},$$

$$\frac{1}{a}\frac{P}{h^2u^2} = 1 - \tfrac{3}{4}k^2 + \tfrac{3}{4}k^2 \cos 2g\theta_1 - \tfrac{1}{2}m^2 - \tfrac{3}{2}m^2 \cos(2 - 2m)\,\theta_1$$
$$+ \tfrac{3}{2}m^2 e \cos c\theta_1 + \tfrac{9}{4}m^2 e \cos(2 - 2m - c)\,\theta_1.$$

The other terms of the third order, which arise in the development of the expression, are neglected because the coefficients of $\theta$ are not near unity,

$$\frac{T}{h^2u^3} = -\frac{3m'a'^3 \{1 + e' \cos(m\theta_1 - \zeta)\}^3}{2h^2a^4 (1 + e \cos c\theta_1)^4} \sin(2 - 2m)\,\theta_1$$
$$= -\tfrac{3}{2}m^2 [1 + 3e' \cos(m\theta_1 - \zeta) - 4e \cos c\theta_1] \sin(2 - 2m)\,\theta_1$$
$$= -\tfrac{3}{2}m^2 \sin(2 - 2m)\,\theta_1 + 3m^2 e \sin(2 - 2m - c)\,\theta_1,$$

$$\frac{T}{h^2u^3} \cdot \frac{du}{d\theta} = \tfrac{3}{2}m^2 ae \sin(2 - 2m)\,\theta_1 \sin c\theta_1$$
$$= \tfrac{3}{4}m^2 ae \cos(2 - 2m - c)\,\theta_1;$$

rejecting the other terms of the third order according to (30).

$$\int\frac{T}{h^2u^3}\,d\theta = \tfrac{3}{2}\,\frac{m^2}{2-2m} \cos(2-2m)\,\theta_1 - \frac{3m^2e}{2 - 2m - c} \cos(2 - 2m - c)\theta_1$$
$$= \tfrac{3}{4}m^2 \cos(2 - 2m)\,\theta_1 - 3m^2 e \cos(2 - 2m - c)\,\theta_1,$$

$$2\left(\frac{d^2u}{d\theta^2} + u\right)\int\frac{T}{h^2u^3}\,d\theta = \tfrac{3}{2}m^2 a \cos(2-2m)\,\theta_1 - 6m^2 ae \cos(2-2m-c)\theta_1.$$

The equation in $u$ becomes

$$\frac{1}{a}\left(\frac{d^2u}{d\theta^2} + u\right) =$$
$$1 - \tfrac{3}{4}k^2 - \tfrac{1}{2}m^2$$
$$+ \tfrac{3}{2}m^2 e \cos c\theta_1$$
$$+ \tfrac{3}{4}k^2 \cos 2g\theta_1$$
$$- 3m^2 \cos(2 - 2m)\,\theta_1$$
$$+ \tfrac{15}{2}m^2 e \cos(2 - 2m - c)\,\theta_1.$$

To integrate this, assume

$$\frac{u}{a} =$$
$$1 + A$$
$$+ e \cos c\theta_1$$
$$+ B \cos 2g\theta_1$$
$$+ C \cos(2 - 2m)\,\theta_1$$
$$+ D \cos(2 - 2m - c)\,\theta_1.$$

Substitute, and equate the coefficients of like terms:

$$A = -\tfrac{3}{4}k^2 - \tfrac{1}{2}m^2, \qquad A = -\tfrac{3}{4}k^2 - \tfrac{1}{2}m^2,$$

$$e(1-c^2) = \tfrac{3}{2}m^2 e, \qquad c = (1-\tfrac{3}{2}m^2)^{\frac{1}{2}} = 1 - \tfrac{3}{4}m^2,$$

$$B(1-4g^2) = \tfrac{3}{4}k^2, \qquad B = \tfrac{3}{4}\frac{k^2}{1-4} = -\tfrac{1}{4}k^2,$$

$$C\{1-(2-2m)^2\} = -3m^2, \qquad C = -\frac{3m^2}{1-4} = +m^2,$$

$$D\{1-(2-2m-c)^2\} = \tfrac{15}{2}m^2 e. \qquad D = \frac{15m^2 e}{2\{1-(1-2m+\frac{3}{4}m^2)^2\}} = \tfrac{15}{8}me.$$

Therefore

$$u = a\left\{\begin{array}{l} 1 - \frac{3}{4}k^2 - \frac{1}{2}m^2 \\ + e\cos(c\theta-\alpha) \\ -\frac{1}{4}k^2\cos 2(g\theta-\gamma) \\ + m^2\cos\{(2-2m)\theta - 2\beta\} \\ + \frac{15}{8}me\cos\{(2-2m-c)\theta - 2\beta + \alpha\} \end{array}\right\} \ldots\ldots U.$$

This is the accurate value of $u$ to the second order.

*To find t to the second order.*

53. The equation to be considered is

$$\frac{dt}{d\theta} = \frac{1}{hu^2}\left(1 + 2\int\frac{T}{h^2u^3}\,d\theta\right)^{-\frac{1}{2}};$$

and, making use of the values of $s$ and $u$ obtained to the second order, we must retrace our steps to recompute $u$ and $\frac{T}{h^2u^3}$ so as to include those terms of the third order in $u$ in which the coefficient of the argument is near zero, and similar terms to the fourth order in $\frac{T}{h^2u^3}$ Art. (48). To do this we need not repeat the whole work, but only so much of it as will enable us to pick out the additional terms in question.

$$\frac{1}{a}\frac{P}{h^2u^2} = 1 - \tfrac{3}{8}\{k\sin g\theta_1 + \tfrac{3}{8}mk\sin(2-2m-g)\theta_1\}^2$$
$$-\tfrac{1}{2}m^2\{1+3e'\cos(m\theta_1-\zeta)-3e\cos c\theta_1\}\{1+3\cos(2-2m)\theta_1\}$$
$$= 1\ldots-\tfrac{9}{8}mk^2\sin g\theta_1\sin(2-2m-g)\theta_1 - \tfrac{3}{2}m^2e'\cos(m\theta_1-\zeta)$$
$$= 1\ldots-\tfrac{9}{16}mk^2\cos(2-2m-2g)\theta_1 - \tfrac{3}{2}m^2e'\cos(m\theta_1-\zeta),$$

$$\frac{T}{h^2u^3} = -\frac{3m'a'^3(1+e'\cos m\theta_1-\zeta)^3\sin(2-2m)\theta_1}{2h^2a^4(1+e\cos c\theta_1-\tfrac{1}{4}k^2\cos 2g\theta_1)^4}$$
$$= -\tfrac{3}{2}m^2\sin(2-2m)\theta_1\{1+e\cos c\theta_1-\tfrac{1}{4}k^2\cos 2g\theta_1\}^{-4}$$
$$= -\tfrac{3}{2}m^2\sin(2-2m)\theta_1\{(1+e\cos c\theta_1)^{-4}+k^2\cos 2g\theta_1\}$$
$$= -\tfrac{3}{2}m^2\sin(2-2m)\theta_1\{\ldots 10e^2\cos^2 c\theta_1+k^2\cos 2g\theta_1\}$$
$$= -\tfrac{3}{2}m^2\sin(2-2m)\theta_1\{\ldots 5e^2\cos 2c\theta_1+k^2\cos 2g\theta_1\}$$
$$= -\tfrac{15}{4}m^2e^2\sin(2-2m-2c)\theta_1 - \tfrac{3}{4}m^2k^2\sin(2-2m-2g)\theta_1.$$

These are the only important *additional* terms.

$$\int\frac{T}{h^2u^3}d\theta = -\tfrac{15}{8}me^2\cos(2-2m-2c)\theta_1 - \tfrac{3}{8}mk^2\cos(2-2m-2g)\theta_1,$$

because $2-2m-2c = -2m+\tfrac{3}{2}m^2$, and $2-2m-2g = -2m-\tfrac{3}{2}m^2$.

Substituting in the differential equation for $u$, the *new* terms in $\dfrac{T}{h^2u^3}\dfrac{du}{d\theta}$ will be of the fourth order and must be neglected, but $2\left(\dfrac{d^2u}{d\theta^2}+u\right)\displaystyle\int\frac{T}{h^2u^3}d\theta$ will bring in the two terms

$-\tfrac{15}{4}me^2a\cos(2-2m-2c)\theta_1$ and $-\tfrac{3}{4}mk^2a\cos(2-2m-2g)\theta_1$;

and we get, so far as important additional terms are concerned,

$$\frac{1}{a}\left\{\frac{d^2u}{d\theta^2}+u\right\} = 1\ldots+\tfrac{15}{4}me^2\cos(2-2m-2c)\theta_1$$
$$+\left(-\tfrac{9}{16}+\tfrac{3}{4}\right)mk^2\cos(2-2m-2g)\theta_1 - \tfrac{3}{2}m^2e'\cos(m\theta_1-\zeta),$$

and the corresponding terms in $u$ will also be of the third order, viz.

$$\frac{u}{a} = \ldots\tfrac{15}{4}me^2\cos(2-2m-2c)\theta_1 + \tfrac{3}{16}mk^2\cos(2-2m-2g)\theta_1$$
$$-\tfrac{3}{2}m^2e'\cos(m\theta_1-\zeta).$$

The complete value of $u$ therefore, so far as important terms are concerned for determining the value of $t$, is

$$\frac{u}{a} = 1 - \tfrac{3}{4}k^2 - \tfrac{1}{2}m^2 + e\cos c\theta_1 - \tfrac{1}{4}k^2\cos 2g\theta_1$$
$$+ m^2\cos(2-2m)\,\theta_1 + \tfrac{15}{8}me\cos(2-2m-c)\,\theta_1$$
$$+ \tfrac{15}{4}m^2\cos(2-2m-2c)\,\theta_1 + \tfrac{3}{16}mk^2\cos(2-2m-2g)\,\theta_1$$
$$- \tfrac{3}{2}m^2e'\cos(m\theta_1 - \zeta),$$

and the complete value of $\int \frac{T}{h^2u^3}\,d\theta$ will be

$$\int \frac{T}{h^2u^3}\,d\theta = \tfrac{3}{4}m^2\cos(2-2m)\,\theta_1 - \tfrac{15}{8}m^2\cos(2-2m-2c)\,\theta_1$$
$$- \tfrac{3}{8}mk^2\cos(2-2m-2g)\,\theta_1,$$

the other term of the third order has no influence on $t$, the coefficient of $\theta$ not being small.

54. We are now in a position to find the value of $t$ to the second order:

$$\frac{dt}{d\theta} = \frac{1}{hu^2}\left(1 + 2\int \frac{T}{h^2u^3}\,d\theta\right)^{-\frac{1}{2}} = \frac{1}{hu^2}\left(1 - \int \frac{T}{h^2u^3}\,d\theta\right).$$

If $\frac{u}{a} = 1 + e\cos c\theta_1 + W$ where $W$ contains all terms of the second and third orders in the expression above, then

$$\frac{1}{hu^2} = \frac{1}{ha^2}(1 + e\cos c\theta_1 + W)^{-2}$$

$$= \frac{1}{ha^2}\{1 - 2(e\cos c\theta_1 + W) + 3(e\cos c\theta_1 + W)^2 - \ldots\}$$

$$= \frac{1}{ha^2}(1 - 2e\cos c\theta_1 - 2W + 3e^2\cos^2 c\theta_1 + 6eW\cos c\theta_1)\ldots$$

$$= \frac{1}{ha^2}(1 + \tfrac{3}{2}e^2 - 2e\cos c\theta_1 - 2W + \tfrac{3}{2}e^2\cos 2c\theta_1 + 6eW\cos c\theta_1).$$

The only part of $6eW\cos c\theta_1$ which gives a term of the third order in which the coefficient of the argument is small, is

$$6e\cos c\theta_1\,\{\ldots+\tfrac{15}{8}me\cos(2-2m-c)\,\theta_1\ldots\},$$

of which one term is

$$+\tfrac{45}{8}me^2\cos(2-2m-2c)\,\theta_1.$$

Therefore $\dfrac{1}{hu^3}$ is equal to

$$\frac{1}{ha^2}\left\{\begin{array}{l}1+\frac{3}{2}e^2+\frac{3}{2}k^2+m^2-2e\cos c\theta_1+\frac{3}{2}e^2\cos 2c\theta_1\\ \quad+\frac{1}{2}k^2\cos 2g\theta_1-2m^2\cos(2-2m)\,\theta_1\\ \quad-\frac{15}{4}me\cos(2-2m-c)\,\theta_1+3m^2e'\cos(m\theta_1-\zeta)\\ \quad-\frac{15}{8}me^2\cos(2-2m-2c)\,\theta_1-\frac{3}{8}mk^2\cos(2-2m-2g)\,\theta_1.\end{array}\right.$$

Also from above

$$1-\int\frac{T}{h^2u^3}\,d\theta=\left\{\begin{array}{l}1-\frac{3}{4}m^2\cos(2-2m)\,\theta_1+\frac{15}{8}me^2\cos(2-2m-2c)\,\theta_1\\ \quad+\frac{3}{8}mk^2\cos(2-2m-2g)\,\theta_1.\end{array}\right.$$

We have now to multiply these results together, and we see that the terms having for coefficients $\frac{15}{8}me^2$ and $\frac{3}{8}mk^2$ will disappear in the product.* These were the terms which were originally of the fourth order in $\dfrac{T}{h^2u^3}$.

$$\frac{dt}{d\theta}=\frac{1}{ha^2}\left\{\begin{array}{l}1+\frac{3}{2}e^2+\frac{3}{2}k^2+m^2-2e\cos c\theta_1+\frac{3}{2}e^2\cos 2c\theta_1\\ \quad+\frac{1}{2}k^2\cos 2g\theta_1-\frac{11}{4}m^2\cos(2-2m)\,\theta_1\\ \quad-\frac{15}{4}me\cos(2-2m-c)\,\theta_1+3m^2e'\cos(m\theta_1-\zeta).\end{array}\right.$$

---

* The principle of the superposition of small disturbances might have led us *à priori* to anticipate the disappearance of these terms:—for, when approximating to the second order we may, according to this principle, calculate separately the effect of the disturbing action of the sun, which is of the second order, and that of any other *independent* disturbance of the same or of a higher order, and then add the results. No term therefore which bears a trace of the action of both these forces should present itself to the second order.

Now since neither $k$ nor $e$ enters into the argument of any term, it is obvious that a term whose coefficient contains $k^2$ or $e^2$ at any stage will retain them through all subsequent operations; and if the terms having $(2-2m-2g)\,\theta_1$ or $(2-2m-2c)\,\theta_1$ for argument could remain in the result to the second order

Let $$\frac{1}{ha^2}(1+\tfrac{3}{2}e^2+\tfrac{3}{2}k^2+m^2)=\frac{1}{p},$$

therefore $p = ha^2(1-\tfrac{3}{2}e^2-\tfrac{3}{2}k^2-m^2)$ to the third order; therefore, multiplying by $p$ and integrating, we get, still to the second order,

$$\left.\begin{aligned} pt=\theta-2e\sin(c\theta-\alpha)&+\tfrac{3}{4}e^2\sin 2(c\theta-\alpha)\\ &+\tfrac{1}{4}k^2\sin 2(g\theta-\gamma)\\ &-\tfrac{11}{8}m^2\sin\{(2-2m)\theta-2\beta\}\\ &-\tfrac{15}{4}me\sin\{(2-2m-c)\theta-2\beta+\alpha\}\\ &+3me'\sin(m\theta+\beta-\zeta)\end{aligned}\right\}\ldots\Theta_2,$$

no constant is added, the time being reckoned from the instant when the mean value of $\theta$ vanishes, for the reasons explained in (Art. 36).

*Coordinates expressed in terms of the time.*

55. The preceding equations $U_2$, $S_2$, $\Theta_2$ give the reciprocal of the radius vector, the latitude and the time in terms of the true longitude; but the principal object of the analytical investigations of the Lunar Theory being the formation of tables which give the coordinates of the moon at stated times, we must express $u$, $s$, and $\theta$ in terms of $t$.

To do this, we must reverse the series $pt=\theta-$ &c., and then substitute the value of $\theta$ in the expressions for $u$ and $s$.

Now $$\begin{aligned}\theta &= pt+2e\sin(c\theta-\alpha) \text{ to the first order}\\ &= pt+2e\sin(cpt-\alpha)\ldots\ldots\ldots\ldots\ldots\ldots;\end{aligned}$$

---

they would there have $k^2$ or $e^2$ respectively in the coefficient, since they have them when they first appear. But $k^2$, $e^2$ and $m$ being independent, such terms must have arisen from the combination of others which originally (in the expressions of the forces) involved these quantities separately:—terms like $k^2\cos 2(g\theta-\gamma)$ and $e^2\cos 2(c\theta-\alpha)$ combined with terms involving $m$: that is, terms which are representatives of disturbing forces of the second order (see note p. 31) combined with others depending on the sun's action. Such terms must therefore be of an order beyond the second.

therefore $c\theta - \alpha = cpt - \alpha + 2e \sin(cpt - \alpha)$ to the first order,

$$2e \sin(c\theta - \alpha) = 2e \{\sin(cpt - \alpha) + 2e \sin(cpt - \alpha) \cos(cpt - \alpha)\}$$

to the second order,

$$= 2e \sin(cpt - \alpha) + 2e^2 \sin 2(cpt - \alpha) \dots\dots\dots\dots ;$$

and as $\theta$ and $pt$ differ by a quantity of the first order, they may be used indiscriminately in terms of the second order; therefere

$$\left.\begin{aligned}\theta = pt + 2e \sin(cpt - \alpha) + \tfrac{5}{4}e^2 \sin 2(cpt - \alpha) \\ - \tfrac{1}{4}k^2 \sin 2(gpt - \gamma) \\ + \tfrac{11}{8}m^2 \sin\{(2 - 2m)pt - 2\beta\} \\ + \tfrac{15}{4}me \sin\{(2 - 2m - c)pt - 2\beta + \alpha\} \\ - 3me' \sin(mpt + \beta - \zeta)\end{aligned}\right\} \dots \Theta'_2.$$

56. In the value of $u$ given in Art. (52), substitute $pt$ for $\theta$ in terms of the second order, and $pt + 2e \sin(cpt - \alpha)$ in the term of the first order; then

$$u = a\left\{\begin{aligned}1 - \tfrac{3}{4}k^2 - \tfrac{1}{2}m^2 - e^2 + e\cos(cpt - \alpha) + e^2 \cos 2(cpt - \alpha) \\ - \tfrac{1}{4}k^2 \cos 2(gpt - \gamma) \\ + m^2 \cos\{(2 - 2m)pt - 2\beta\} \\ + \tfrac{15}{8}me \cos\{(2 - 2m - c)pt - 2\beta + \alpha\}\end{aligned}\right\} \dots U'_2.$$

57. Similarly, the expression for $s$ becomes

$$s = k \sin\{(gpt - \gamma) + 2e \sin(cpt - \alpha)\}$$
$$+ \tfrac{3}{8}mk \sin\{(2 - 2m - g)pt - 2\beta + \gamma\};$$

or
$$s = k\left\{\begin{aligned}&\sin(gpt - \gamma) \\ &+ e \sin\{(g + c)pt - \alpha - \gamma\} \\ &- e \sin\{(g - c)pt + \alpha - \gamma\} \\ &+ \tfrac{3}{8}m \sin\{(2 - 2m - g)pt - 2\beta + \gamma\}\end{aligned}\right\} \dots\dots S'_2.$$

The expression for $s$ is more complex in this form than when given in terms of the true longitude $\theta$.

*Moon's Parallax.*

58. If $P$ be the moon's *mean* parallax, and $\Pi$ the parallax at the time $t$,

$$\Pi = \frac{\text{radius of earth}}{\text{distance of } ☽} = \frac{R}{\frac{1}{u}\sqrt{(1+s^2)}} = Ru\left(1-\tfrac{1}{2}s^2\right) \text{ to the third order,}$$

$$= Ru\left\{1-\tfrac{1}{4}k^2+\tfrac{1}{4}k^2\cos 2\left(gpt-\gamma\right)\right\} \text{ to the second order,}$$

$$= Ra\begin{cases}1-k^2-\tfrac{1}{2}m^2-e^2+e\cos(cpt-\alpha)+e^2\cos 2\,(cpt-\alpha)\\ \qquad +m^2\cos\{(2-2m)\,pt-2\beta\}\\ \qquad +\tfrac{15}{8}me\cos\{(2-2m-c)\,pt-2\beta+\alpha\};\end{cases}$$

but $P$ = the portion which is independent of periodical terms,

$$= Ra\left(1-k^2-\tfrac{1}{2}m^2-e^2\right);$$

$$\text{therefore } \Pi = P\begin{cases}1+e\cos(cpt-\alpha)+e^2\cos 2\,(cpt-\alpha)\\ +m^2\cos\{(2-2m)\,pt-2\beta\}\\ +\tfrac{15}{8}me\cos\{(2-2m-c)\,pt-2\beta+\alpha\}\end{cases}$$

neglecting terms of the third order. We see that, to the second order, the variable part of the parallax is independent of the inclination.

59. Here we terminate our approximations to the values of $u$, $s$, and $\theta$. If we wished to carry them to the third order, it would be necessary to include some terms of the fourth and fifth orders according to Art .(30), and the approximate values of $P$, $T$, and $S$, given in Art. (24), would no longer be sufficiently accurate, but wc should have to recur to the exact values, and from them obtain terms of an order beyond those already employed.* The process followed in the preceding pages is a sufficient clue to what would have to be done for a higher approximation.

* See *Parallactic Inequality*, Art. (105).

The coordinates $u'$ and $\theta'$ of the sun's position are, by the theory of elliptic motion, known in terms of the time $t$, and $t$ is given in terms of the longitude $\theta$ by the equation $\Theta_2$. Hence $u'$ and $\theta'$ can be obtained in terms of $\theta$; but it will be necessary to take into account the slow progressive motion of the sun's perigee, which we have hitherto neglected. This we may do by writing $c'\theta' - \zeta$ for $\theta' - \zeta$, $c'$ being a quantity which differs very little from unity.*

These values of $u'$, $\theta'$, together with those of $u$ and $s$ given by $U_2$ and $S_2$, are then to be substituted in the corrected values of the forces, and thence in the differential equations. The integrations being performed as before will give the values of $u$, $s$, and $t$ in terms of $\theta$ to the third order, and from these, as in Arts. (55), (56), and (57), may be obtained $u$, $s$, and $\theta$ in terms of $t$.

60. More approximate values of $c$ and $g$ are obtained at the same time, by means of the coefficients of cos $(c\theta - \alpha)$ and sin $(g\theta - \gamma)$ in the differential equations, (see Appendix, Arts. 102 and 103).

61. The values to the fourth order are then obtained from those to the third by continuing the same process, and

---

* 'En réfléchissant sur les termes que doivent introduire toutes les quantités 'précédentes, on voit qu'il se peut glisser des cosinus de l'angle $\theta$ dont nous avons 'vu le dangereux effet d'amener dans la valeur de $u$ des arcs au lieu de leurs 'cosinus; de tels termes viendront, par exemple, de la combinaison des cosinus 'de $(1-m)\,\theta$ avec des cosinus de $m\theta$.........

'......... Pour éviter cet inconvenient qui ôterait à la solution précédente 'l'avantage de convenir à un aussi grand nombre de révolutions qu'on voudrait, 'et la priverait de la simplicité et de l' universalité si précieuses en mathématiqnes, 'il faut commencer par en chercher la cause. Or, on découvre facilement que 'ces termes ne viennent que de ce qu'on a supposé fixe l'apogée du soleil, ce 'qui n'est pas permis en toute rigueur, puisque quelque petite qne soit sur cet astre 'l'action de la lune, elle n'en est pas moins réelle et doit lui produire un mouvement 'd'apogée quoique très lent à la vérité'.—Clairaut, *Theorie de la Lune*, p. 55. 2me Edition.

so on to the fifth and higher orders; but the calculations are so complex that the approximations have not been carried beyond the fifth order, and already the value of $\theta$ in terms of $t$ contains 128 periodical terms, without including those due to the disturbances produced by the planets. The coefficients of these periodical terms are functions of $m$, $e$, $e'$, $\frac{a'}{a}$, $c$, $g$, $k$, and are themselves very complicated under their literal forms: that of the term whose argument is twice the difference of the longitude of the sun and moon, for instance, is itself composed of 46 terms, combinations of the preceding constants.

See Pontécoulant, *Système du Monde*, tom. IV., p. 572.

# CHAPTER V.

## NUMERICAL VALUES OF THE COEFFICIENTS.

62. Having thus, from theory, obtained the *form* of the developments of the coordinates of the moon's position at any time, the next necessary step is the determination of the numerical values of the coefficients of the several terms.

We here give three different methods which may be employed for that purpose, and these may, moreover, be combined according to circumstances.

63. *First Method.* The values of the constants $p$, $m$, $\alpha$, $\beta$, $\gamma$, $\zeta$ which enter into the *arguments*, and of the additional ones $a$, $k$, $e$, &c. which enter into the *coefficients*, of the terms in the previous developments, may be obtained with great accuracy from observation.

For this we must employ observations separated by very long intervals, such, for instance, as ancient and modern eclipses; also particular observations of the sun and moon made when the bodies occupy certain selected positions; and other observations of a special character.

When the values of the elements have been so obtained, then the theoretical values of the coefficients may be computed by substitution in the analytical expressions.

64. *Second Method.* Let the constants which enter into the *arguments* be determined as in the first method; and let a large number of observations be made, from each of which a value of the true longitude, latitude, or parallax is obtained,

together with the corresponding value of $t$ reckoned from the fixed epoch when the mean longitude is zero. Let these corresponding values be substituted in the equations, each observation thus giving rise to a relation between the unknown constant *coefficients*.

A very great number of equations having been thus obtained, let them, by the method of least squares or some analogous process, be reduced to as many as there are coefficients to be determined. The solution of these simple equations will give the required values.

This method, however, would scarcely be practicable in a high order of approximation. For instance, in the fifth order, as stated in Art. (61), each observation would give rise to an equation containing 129 unknown quantities and the immense number of equations so obtained would have to be reduced to 129 equations of 130 terms each.

65. *Third Method.* When the constants which enter into the arguments have been determined by the first method, we may obtain any one of the coefficients independently of all the others by the following process, provided the number of observations be very great.

Let the form of the function be

$$V = A + B\sin\theta + C\sin\phi + \&c.,$$

and let it be required to determine the constants $A$, $B$, $C$, &c. separately; $\theta$, $\phi$, &c., being known functions of the time.

Let the results of a great number of observations corresponding to values $\theta_1$, $\theta_2$, $\theta_3$, &c., $\phi_1$, $\phi_2$, $\phi_3$, &c., be $V_1$, $V_2$, $V_3$, &c.; so that

$$\begin{aligned} V_1 &= A + B\sin\theta_1 + C\sin\phi_1 + \&c., \\ V_2 &= A + B\sin\theta_2 + C\sin\phi_2 + \&c., \\ V_3 &= A + B\sin\theta_3 + C\sin\phi_3 + \&c., \\ &\vdots \\ V_n &= A + B\sin\theta_n + C\sin\phi_n + \&c. \end{aligned}$$

Now, $n$ being *very great*, we may assume that the sum of the positive values of each periodical term will be about counterbalanced by the sum of its negative values; and therefore, that if we add all the equations together these terms will disappear;

therefore $$A = \frac{V_1 + V_2 + V_3 + \ldots\ldots + V_n}{n},$$

which determines the non-periodic part of the function.

To determine $B$. Let the observations be divided into two sets separating the positive and negative values of $\sin\theta$; then the other periodical terms, not having the same period, may be considered as cancelling themselves in adding up the terms of each set. Let there be $r$ terms in the first set and $s$ terms in the second, and let $V'$, $V''$, ...... $V^r$ be the values of $V$ corresponding to positive values of $\sin\theta$, which values we may assume to be uniformly distributed from $\sin\theta$ to $\sin\pi$, and therefore to be $\sin\delta\theta$, $\sin 2\delta\theta$, ...... $\sin r.\delta\theta$, where $r.\delta\theta = \pi$.

And, again, let $V_{,}$, $V_{,,}$, $V_{,,,}$, ...... $V_s$, be the values of $V$ corresponding to the negative values of $\sin\theta$; viz., $\sin(-\Delta\theta)$, $\sin(-2\Delta\theta)$, ...... $\sin(-s.\Delta\theta)$, where $s.\Delta\theta = \pi$. Then,

$$V' = A + B\sin\delta\theta + C\sin\phi' + \ldots,$$
$$V'' = A + B\sin 2.\delta\theta + C\sin\phi'' + \ldots,$$
$$\vdots$$
$$V^r = A + B\sin r.\delta\theta + C\sin\phi^r + \ldots;$$

therefore

$$V' + V'' + \ldots + V^r = r.A + B\Sigma_0^\pi(\sin\theta)$$
$$= rA + \frac{B}{\delta\theta}\Sigma_0^\pi(\sin\theta.\delta\theta);$$

therefore

$$\frac{V' + V'' + \ldots + V^r}{r} = A + \frac{B}{\pi}\int_0^\pi \sin\theta.d\theta$$
$$= A + \frac{2B}{\pi}.$$

$$V_{,} = A - B\sin\Delta\theta + C\sin\phi_{,} + \ldots,$$
$$V_{,,} = A - B\sin 2.\Delta\theta + C\sin\phi_{,,} + \ldots,$$
$$\vdots$$
$$V_s = A - B\sin s.\Delta\theta + C\sin\phi_s + \ldots;$$

therefore

$$V_{,} + V_{,,} + \ldots + V_s = s.A - B\Sigma_0^\pi(\sin\theta)$$
$$= sA - \frac{B}{\Delta\theta}\Sigma_0^\pi(\sin\theta.\Delta\theta);$$

therefore

$$\frac{V_{,} + V_{,,} + \ldots + V_s}{s} = A - \frac{B}{\pi}\int_0^\pi \sin\theta.d\theta$$
$$= A - \frac{2B}{\pi};$$

$$\text{therefore } B = \frac{\pi}{4}\left(\frac{V' + V'' + \ldots + V^r}{r} - \frac{V_{,} + V_{,,} + \ldots + V_s}{s}\right);$$

and in a similar manner may each of the coefficients be independently determined.*

66. When the periods of two of the terms differ but slightly—for instance if $\theta$ and $\phi$ go through their periodic variations very nearly in the same time,—the method could not then with safety be applied; for, since the same values of $\theta$ and $\phi$ would very nearly recur together during a longer time than that through which the observations would extend, the two terms would be so blended in the value of $V$ that they would enter nearly as one term—the difference between $\theta$ and $\phi$ would be very nearly the same at the end as at the beginning of the series of observations.

---

* If $r$ and $s$ are not sufficiently great to allow us to substitute $\int_0^\pi \sin\theta\, d\theta$ for $\Sigma_0^\pi \sin\theta \,.\, \delta\theta$, we must proceed as follows:

$$\begin{aligned} V' + V'' + \ldots\ldots + V^r &= rA + B\,(\sin\delta\theta + \sin 2\delta\theta + \ldots\ldots + \sin r\delta\theta) \\ &= rA + B\,\frac{\sin\frac{1}{2}(r+1)\,\delta\theta \sin\frac{1}{2}r\,.\,\delta\theta}{\sin\frac{1}{2}\delta\theta} \\ &= rA + B\,\frac{\cos\frac{1}{2}\delta\theta}{\sin\frac{1}{2}\delta\theta}, \end{aligned}$$

$$\begin{aligned} \frac{V' + V'' + \ldots\ldots + V^r}{r} &= A + \frac{B}{r}\,\frac{1 - \frac{1}{8}\delta\theta^2}{\frac{1}{2}\delta\theta - \frac{1}{48}\delta\theta^3}, \text{ nearly}, \\ &= A + \frac{2B}{\pi}\,\frac{1 - \frac{1}{8}\dfrac{\pi^2}{r^2}}{1 - \frac{1}{24}\dfrac{\pi^2}{r^2}} \\ &= A + \frac{2B}{\pi}\left(1 - \frac{1}{12}\,\frac{\pi^2}{r^2}\right). \end{aligned}$$

Similarly, $$\frac{V_{,} + V_{,,} + \ldots\ldots + V_s}{s} = A - \frac{2B}{\pi}\left(1 - \frac{1}{12}\,\frac{\pi^2}{s^2}\right);$$

therefore

$$\begin{aligned} B &= \frac{\pi}{4\left\{1 - \dfrac{\pi^2}{24}\left(\dfrac{1}{r^2} + \dfrac{1}{s^2}\right)\right\}}\left(\frac{V' + V'' + \ldots\ldots + V^r}{r} - \frac{V_{,} + V_{,,} + \ldots\ldots + V_s}{s}\right) \\ &= \frac{\pi}{4}\left\{1 + \frac{\pi^2}{24}\left(\frac{1}{r^2} + \frac{1}{s^2}\right)\right\}\left(\frac{V' + V'' + \ldots\ldots + V^r}{r} - \frac{V_{,} + V_{,,} + \ldots\ldots + V_s}{s}\right). \end{aligned}$$

67. Let us suppose the periods to be actually identical, so that $\phi = \theta + \alpha$, $\alpha$ being some constant angle; then

$$B \sin\theta + C \sin\phi$$

may be written $(B + C\cos\alpha)\sin\theta + C\sin\alpha\cos\theta$,

or $$V = A + (B + C\cos\alpha)\sin\theta + C\sin\alpha\cos\theta + \ldots\ldots.$$

If now we divide the observatious, as before, into two sets, corresponding to the positive and negative values of $\sin\theta$, the terms involving $\cos\theta$ will disappear in the summation of each set; and, following the process of the method, we shall find the value of

$$B + C\cos\alpha = M \text{ suppose.}$$

Dividing again into two sets corresponding to the positive and negative values of $\cos\theta$, the terms in $\sin\theta$ will be cancelled, and the same process will give the value of

$$C\sin\alpha = N \text{ suppose.}$$

Treating the observations in the same way with respect to the angle $\phi$, we get two results,

$$C + B\cos\alpha = M',$$

$$-B\sin\alpha = N';$$

from these four equations we easily get

$$B = \frac{M'N + MN'}{M^2 - M'^2} N' \text{ or } = -\frac{M'N + MN'}{N^2 - N'^2} N',$$

$$C = -\frac{N}{N'} B;$$

$M, N, M', N'$ are connected by the equation of condition,

$$M^2 - M'^2 = N'^2 - N^2.$$

68. When the periods of $\theta$ and $\phi$ are nearly, but not exactly, the same, this equation of condition will not hold, and the preceding values of $B$ and $C$ would not be exactly

correct; but yet they would be very approximate, especially if the mean between the two values of $B$ be taken.

Or, we may, after having taken one of these slightly erroneous values for $B$, make a further correction by establishing as it were a counterbalancing error in the value of $C$. Let $B'$ be the value so found for $B$; then, from the $V$ of each of the observations subtract the value $B' \sin \theta$, the result $U$ will be very nearly equal to $A + C \sin \phi +$ &c., and from the $n$ equations

$$U_1 = A + C \sin \phi_1 + \ldots\ldots\ldots\ldots$$
$$U_2 = A + C \sin \phi_2 + \ldots\ldots\ldots\ldots$$
$$U_n = A + C \sin \phi_n + \ldots\ldots\ldots\ldots$$

a value $C'$ of $C$ will be obtained, by the rule, which will be very approximate, and, at the same time, agree better with $B'$ in satisfying the equations than $C$ itself would do.

69. When two terms whose periods are nearly equal do occur, it is plain, by examining the values of $M$ and $M'$, that the errors which would be committed by following the rule, without taking account of this peculiarity, would be the taking $B + C \cos \alpha$ and $C + B \cos \alpha$ for $B$ and $C$ respectively.

# CHAPTER VI.

## PHYSICAL INTERPRETATION.

70. The solution of the problem which is the object of the Lunar Theory may now be considered as effected; that is, we have obtained equations which enable us to assign the moon's position in the heavens at any given time to the second order of approximation; we have explained how the numerical values of the coefficients in these equations may be determined from observation; and we have, moreover, shewn how to proceed in order to obtain a higher approximation.*

It will, however, be interesting to discuss the results we have arrived at:—to see whether they will enable us to form some idea of the nature of the moon's complex motion; also how far they will explain those inequalities or departures from uniform circular motion which ancient astronomers had observed; but which, until the time of Newton, were so many unconnected phenomena; or, at least, had only such arbitrary connexions as the astronomers chose to assign, by grafting one eccentric or epicycle on another as each newly discovered inequality seemed to render it necessary.

It is true that our expressions, composed of periodic terms, are nothing more than translations into analytical

* The means of taking into account the ellipsoidal figure of the earth and the disturbances produced by the planets, are too complex to form part of an introductory treatise. For information on these points reference may be made to Airy's *Figure of the Earth*. Pontécoulant's *Système du Monde*, vol. IV.

language of the epicycles of the ancient;* but they are evolved directly from the fundamental laws of force and motion, and as many new terms as we please may be obtained by carrying on the same process; whereas the epicycles of Hipparchus and his followers were the result of numerous and laborious observations and comparisons of observations; each epicycle being introduced to correct its predecessor when this one was found inadequate to give the position of the body at all times: just as with us, the terms of the second order correct the rough results given by those of the first; the terms of the third order correct those of the second, and so on. But it is impossible to conceive that observation alone could have detected *all* those minute irregularities which theory makes known to us in the terms of the third and higher orders, even supposing our instruments far more perfect than they are; and it will always be a subject of admiration and surprise, that Tycho, Kepler, and their predecessors should have been able to *feel* their way so far among the Lunar inequalities, with the means of observation they possessed.

## LONGITUDE OF THE MOON.

71. We shall first discuss the expression for the moon's longitude, as found Art. (55),

$$\begin{aligned}\theta = pt &+ 2e\sin(cpt-\alpha) + \tfrac{5}{4}e^2\sin 2(cpt-\alpha)\\ &+ \tfrac{15}{4}me\sin\{(2-2m-c)pt-2\beta+\alpha\}\\ &+ \tfrac{11}{8}m^2\sin\{(2-2m)pt-2\beta\}\\ &- 3me'\sin(mpt+\beta-\zeta)\\ &- \tfrac{1}{4}k^2\sin 2(gpt-\gamma).\end{aligned}$$

The mean value of $\theta$ is $pt$; and in order to judge of the effect of any of the small terms, we may consider them

---

* See Whewell's *History of the Inductive Sciences.*

one at a time as a correction on this mean value $pt$, or we may select a combination of two or more to form this correction.

We shall have instances of combination in the *elliptic inequality* and the *evection*, Arts. (73) and (77); but in the remaining inequalities each term of the expression will form a correction to be considered by itself.

72. Neglecting all the periodical terms, we have

$$\theta = pt,$$

$$\frac{d\theta}{dt} = p,$$

which indicates uniform angular velocity; and as, to the same order, the value of $u$ is constant, the two together indicate that the moon moves uniformly in a circle, the period of a revolution being $\frac{2\pi}{p}$, which is, therefore, the expression for a mean sidereal month, or about $27\frac{1}{3}$ days.*

The value of $p$ is, according to Art. (54), given by

$$\frac{1}{p} = \frac{1}{ha^2}(1 + \tfrac{3}{2}k^2 + m^2 + \tfrac{3}{2}e^2),$$

and as $m$ is due to the disturbing action of the sun, we see that the mean angular velocity is less, and therefore the mean periodic time greater than if there were no disturbance.

### *Elliptic Inequality or Equation of the Centre.*

73. We shall next consider the effect of the first three terms together: the effect of the second alone, as a correction of $pt$, will be discussed in the Historical Chapter, Art. (115).

---

* The accurate value was 27d. 7h. 43m. 11·261s. in the year 1801. See Art. (108).

$$\theta = pt + 2e\sin(cpt - \alpha) + \tfrac{5}{4}e^2 \sin 2(cpt - \alpha),$$

which may be written

$$\theta = pt + 2e\sin[pt - \{\alpha + (1-c)pt\}] + \tfrac{5}{4}e^2 \sin 2[pt - \{\alpha + (1-c)pt\}].$$

But the connexion between the longtitude and the time in an ellipse described about a centre of force in the focus is, Art. (13), to the second order of small quantities:

$$\theta = nt + 2e\sin(nt - \alpha') + \tfrac{5}{4}e^2 \sin 2(nt - \alpha'),$$

where $n$ is the mean motion, $e$ the eccentricity, and $\alpha'$ the longitude of the apse.*

Hence, the terms we are now considering indicate motion in an ellipse; the mean motion being $p$, the eccentricity $e$, and the longitude of the apse $\alpha + (1-c)\,pt$; that is, the apse is not stationary but has a progressive motion in longitude, uniform, and equal to $(1-c)\,p$.

74. The two terms $2e\sin(cpt - \alpha) + \frac{5}{4}e^2 \sin 2(cpt - \alpha)$ constitute the *elliptic inequality*, and their effect may be further illustrated by means of a diagram.

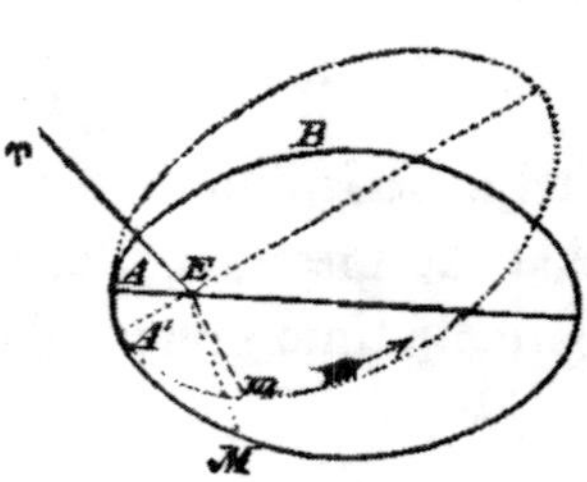


Let the full line $AMB$ represent the moon's orbit about the earth $E$, when the time $t$ commences, that is, when the moon's *mean* place is in the prime radius $E\Upsilon$, from which the longitudes are reckoned.

The angle $\Upsilon EA$, the longitude of the apse, is then $\alpha$. At the time $t$, when the moon's mean longitude is $\Upsilon EM = pt$, the apse line will have moved

---

* The epoch $\epsilon$ which appears in the expression of Art. (13) is here omitted; a proper assumption for the origin of $t$, as explained in Art. (36), enabling us to avoid the $\epsilon$.

in the same direction through the angle $AEA' = (1-c)\,\Upsilon EM$, and the orbit will have taken the position indicated by the dotted ellipse. The true place of the moon in this orbit, so far as these two terms are concerned, will be $m$, where

$$\begin{aligned} MEm &= 2e \sin (cpt - \alpha) + \tfrac{5}{4}e^2 \sin 2\,(cpt - \alpha) \\ &= 2e \sin A'EM \quad + \tfrac{5}{4}e^2 \sin 2A'EM \\ &= 2e \sin A'EM\,(1 + \tfrac{5}{4}e \cos A'EM)\,; \end{aligned}$$

which, since $e$ is about $\frac{1}{20}$, is positive from perigee to apogee, and therefore the *true* place before the *mean;* and the contrary from apogee to perigee: at the apses the places will coincide.

75. The angular velocity of the apse is $(1-c)\,p$, or, if for $c$ we put the value found in Art. (52), the velocity will be $\frac{3}{4}m^2p$. Hence, while the moon describes 360°, the apse should describe $\frac{3}{4}m^2.360° = 1\frac{3}{5}°$ nearly, $m$ being about $\frac{1}{13}$.

But Hipparchus had found, and all modern observations confirm his result, that the motion of the apse is about 3° in each revolution of the moon. See Art. (118).

This difference arises from our value of $c$ not being represented with sufficient accuracy by $1 - \frac{3}{4}m^2$.

Newton himself was aware of this apparent discrepancy between his theory and observation; and we are led, by his own expressions (Scholium to Prop. 35, lib. III. in the first edition of the *Principia*), to conclude that he had got over the difficulty. This is rendered highly probable when we consider that he had solved a somewhat similar problem in the case of the node; but he has nowhere given a statement of his method: and Clairaut, to whom we are indebted for the solution, was on the point of publishing a new hypothesis of the laws of attraction, in order to account for it, when it occurred to him to carry the approximations to the third order, and he there found the next term in the value of $c$ nearly as considerable as the one already obtained. See Appendix. The value is

$$c = 1 - \tfrac{3}{4}m^2 - \tfrac{225}{32}m^3;$$

$$\therefore 1 - c = \tfrac{3}{4}m^2 + \tfrac{225}{32}m^3 = \tfrac{3}{4}m^2\left(1 + \tfrac{75}{8}m\right);$$

$$\therefore (1-c)\,360^\circ = \left(1 + \tfrac{75}{104}\right)\text{(value found previously)}$$

$$= 2\tfrac{3}{4}^\circ \text{ nearly,}$$

thus reconciling theory and observation, and removing what had proved a great stumbling-block in the way of all astronomers.*

When the value of $c$ is carried to higher orders of approximation, the most perfect agreement is obtained.

The motion of the apse line is considered by Newton in his *Principia*, lib. I., Prop. 66, Cor. 7.

### *Evection.*

76. The next term $+\tfrac{15}{4}me\sin\{(2-2m-c)\,pt - 2\beta + \alpha\}$ in the value of $\theta$ has been named the *Evection*. We shall consider its effect in two different ways.

First, by itself, as forming a correction on $pt$,

$$\theta = pt + \tfrac{15}{4}me\sin\{(2-2m-c)\,pt - 2\beta + \alpha\}.$$

Let $☽ = pt$ = moon's mean longitude at time $t$,

$\odot = mpt + \beta$ = sun's ..............................,

$\alpha' = (1-c)\,pt + \alpha$ = mean longitude of apse .........,

then

$$\theta = pt + \tfrac{15}{4}me\sin\left[2\{pt - (mpt+\beta)\} - \{pt - \overline{(1-c)\,pt + \alpha}\}\right]$$

$$= pt + \tfrac{15}{4}me\sin\{2(☽ - \odot) - (☽ - \alpha')\}.$$

---

* See Dr. Whewell's *Bridgewater Treatise*.

The effect of this term will therefore be as follows:

In syzygies

$$\theta = pt - \tfrac{15}{4} me \sin (☽ - \alpha');$$

or the true place of the moon will be before or behind the mean, according as the moon, at the same time, is between apogee and perigee or between perigee and apogee.

In quadratures

$$\theta = pt + \tfrac{15}{4} me \sin (☽ - \alpha'),$$

and the circumstances will be exactly reversed.

In both cases, the correction will vanish when the apse happens to be in syzygy or quadrature with regard to the sun at the same time as the moon.

In intermediate positions, the nature of the correction is more complex, but it will always vanish when the sun is at the middle point between the moon and the apse, or when distant 90° or 180° from that point; for if

$$\odot = \frac{☽ + \alpha'}{2} - r.90°, \text{ where } r = 0, \pm 1, \text{ or } 2,$$

$$\begin{aligned} \sin [2(☽ - \odot) - (☽ - \alpha')] &= \sin (☽ + \alpha' - 2\odot) \\ &= \sin r.180° \\ &= 0. \end{aligned}$$

77. The other and more usual method of considering the effect of this term is in combination with the two terms of the elliptic inequality, as follows:

*To determine the change in the position of the apse and in the eccentricity of the moon's orbit produced by the evection.*

Taking the elliptic inequality and the evection together, we have

$$\theta = pt + 2e \sin (cpt - \alpha) + \tfrac{5}{4} e^2 \sin 2(cpt - \alpha) + \tfrac{15}{4} me \sin \{(2 - 2m - c) pt - 2\beta + \alpha\}.$$

Let $\alpha'$ be the longitude of the apse at time $t$ on supposition of uniform progression,

$\odot$ ............................ sun ...........;

whence
$$\alpha' = (1-c)\,pt + \alpha,$$
$$\odot = mpt + \beta.$$

And the above may be written
$$\theta = pt + 2e\sin(cpt-\alpha) + \tfrac{5}{4}e^2\sin 2(cpt-\alpha) + \tfrac{15}{4}me\sin\{cpt-\alpha+2(\alpha'-\odot)\};$$

and the second and fourth terms may be combined into one,
$$2E\sin(cpt-\alpha+\delta),$$

if we assume $E\cos\delta = e + \frac{15}{8}me\cos 2(\alpha'-\odot)$,

and $E\sin\delta = \frac{15}{8}me\sin 2(\alpha'-\odot)$;

whence $\tan\delta$ and $E$ may be found; and approximately,
$$E = e\{1+\tfrac{15}{8}m\cos 2(\alpha'-\odot)\},$$
$$\delta = \tfrac{15}{8}m\sin 2(\alpha'-\odot).$$

The term $\frac{5}{4}e^2\sin 2(cpt-\alpha)$ may also, therefore, to the second order, be expressed by
$$\tfrac{5}{4}E^2\sin 2(cpt-\alpha+\delta),$$

and the longitude becomes
$$\theta = pt + 2E\sin(cpt-\alpha+\delta) + \tfrac{5}{4}E^2\sin 2(cpt-\alpha+\delta),$$

or $\theta = pt + 2E\sin(pt-\alpha'+\delta) + \frac{5}{4}E^2\sin 2(pt-\alpha'+\delta)$;

but the last two terms constitute elliptic inequality in an orbit whose eccentricity is $E$ and longitude of the apse $\alpha'-\delta$; therefore the evection, taken in conjunction with elliptic inequality, has the effect of rendering the eccentricity of the moon's orbit variable, increasing it by $\frac{15}{8}me$ when the apse-line is in syzygy, and diminishing it by the same quantity when the apse-line is in quadrature; the general expression for the increment being $\frac{15}{8}me\cos 2(\alpha'-\odot)$.

And another effect of this term is, to diminish the longitude of the apse, calculated on the supposition of its uniform progression, by the quantity $\delta = \frac{15}{8} m \sin 2(\alpha' - \odot)$; so that the apse is behind its mean place when in the first and third quadrant in advance of the sun, and before its mean place in the second and fourth.

The cycle of these changes will evidently be completed in the period of half a revolution of the sun with respect to the apse, or in about $\frac{9}{16}$ of a year.*

78. The period of the evection itself, considered independently of its effect on the orbit, is the time in which the argument $(2 - 2m - c)pt - 2\beta + \alpha$ will increase by $2\pi$.

Therefore period of evection

$$= \frac{2\pi}{(2 - 2m - c)p} = \frac{\text{mean sidereal month}}{2 - 2m - c}$$

$$= \frac{\text{mean sidereal month}}{1 - 2m + \frac{3}{4}m^2} = \frac{27\frac{1}{3} \text{ days}}{1 - \frac{2}{13}}, \text{ nearly,}$$

$$= 31\tfrac{1}{2} \text{ days, nearly.}\dagger$$

Newton has considered the evection, so far as it arises from the central disturbing force, in Prop. 66, Cor. 9, of the *Principia*.

## *Variation.*

79. *To explain the physical meaning of the term*

$$\tfrac{11}{8} m^2 \sin \{(2 - 2m)pt - 2\beta\},$$

---

* The change of eccentricity and the variation in the motion of the apse follow the same law as the abscissa and ordinate of an ellipse referred to its centre: for if $E - e = x$ and $\delta = y$, then

$$\frac{x^2}{(\frac{15}{8}me)^2} + \frac{y^2}{(\frac{15}{8}m)^2} = 1.$$

† The accurate value is 31·8119 days.

*in the expression for the moon's longitude,*

$$\theta = pt + \tfrac{11}{8} m^2 \sin\{(2-2m)pt - 2\beta\}.$$

Let ☽ represent the moon's mean longitude at time $t$,

⊙ .................. sun's ......................................

therefore $$☽ = pt,$$
$$⊙ = mpt + \beta;$$

and the value of $\theta$ becomes

$$\theta = pt + \tfrac{11}{8} m^2 \sin 2(☽ - ⊙),$$

which shews that from syzygy to quadrature the moon's true place is before the mean, and behind it from quadrature to syzygy; the maximum difference being $\frac{11}{8}m^2$ in the octants.

The angular velocity of the moon, so far as this term is concerned, is

$$\frac{d\theta}{dt} = p + \tfrac{11}{4}(1-m)m^2 p \cos 2(☽ - ⊙),$$

$$= p\{1 + \tfrac{11}{4} m^2 \cos 2(☽ - ⊙)\}, \text{ nearly},$$

which exceeds the mean angular velocity $p$ at syzygies, is equal to it in the octants, and less in the quadratures.

This inequality has been called the *Variation*, its period is the time in which the argument $(2-2m)pt - 2\beta$ will increase by $2\pi$;

$$\therefore \text{ period of variation} = \frac{2\pi}{(2-2m)p} = \frac{\text{mean synodical month}}{2}$$

$$= 14\tfrac{3}{4} \text{ days, nearly.*}$$

80. The quantity $\frac{11}{8}m^2$ is only the first term of an endless series which constitutes the coefficient of the variation, the other terms being obtained by carrying the approximation

* The accurate value is 14·765294 days.

to a higher order. It is then found that the next term in the coefficient is $\frac{59}{12}m^3$, which is about $\frac{3}{11}$ of the first term; and as there are several other important terms, it is only by carrying the approximation to a higher order (the 5th at least) that the value of this coefficient can be obtained with sufficient accuracy from theory. In fact, $\frac{11}{8}m^2$ would give a coefficient of $26' 27''$ only; whereas the accurate value is found to be $39' 30''$.

The same remark applies also to the coefficients of all the other terms.

81. As far as terms of the second order, the coefficient of the variation is independent of $e$ the eccentricity, and $k$ the inclination of the orbit. It would therefore be the same in an orbit originally circular, whose plane coincided with the plane of the ecliptic: it is thus that Newton has considered it. *Princip.* prop. 66, cor. 3, 4, and 5.

## *Annual Equation.*

82. *To explain the physical meaning of the term*

$$-3me' \sin(mpt + \beta - \zeta)$$

*in the expression for the moon's longitude.*

$\theta = pt - 3me' \sin(mpt + \beta - \zeta)$,

$\quad = pt - 3me' \sin$ (longitude of sun − longitude of sun's perigee),

$\quad = pt - 3me' \sin$ (sun's anomaly).

Hence, while the sun moves from his perigee to his apogee, the true place of the moon will be behind the mean; and from apogee to perigee, before it. The period being an anomalistic year, the effect is called *Annual Equation.*

Differentiating $\theta$ we get

$$\frac{d\theta}{dt} = p\{1 - 3m^2e' \cos(\text{sun's anomaly})\}.$$

Hence, so far as this inequality is concerned, the moon's angular velocity is least when the sun is in perigee, that is *at present* about the 1st of January, and greatest when the sun is in apogee, or about the 1st of July.

The annual equation is, to this order, independent of the eccentricity and inclination of the moon's orbit, and therefore, like the variation, would be the same in an orbit originally circular. *Vide* Newton, *Principia*, prop. 66, cor. 6.

*Reduction.*

83. Before considering the effect of the term

$$-\frac{k^2}{4}\sin 2(gpt-\gamma),$$

which, as we shall see Art. (89), is very nearly equal to the difference between the longitude in the orbit and the longitude in the ecliptic, it will be convenient to examine the expression for the latitude of the moon, and to see how the motion of the node is connected with the value of $g$.

## LATITUDE OF THE MOON.

84. The expression found for the tangent of the latitude,* Art. (51), is

$$s = k\sin(g\theta-\gamma) + \tfrac{3}{8}mk\sin\{(2-2m-g)\,\theta-2\beta+\gamma\}.$$

If we reject all small terms, we have

$$s=0,$$

or the orbit of the moon coinciding with the ecliptic, which is a first rough approximation to its true position.

---

* This expression for the tangent of the latitude is more convenient than that which gives it in terms of the mean longitude, Art. (57), on account of the less number of terms involved. See Pontécoulant, vol. IV. p. 630.

85. Taking the first term of the expansion

$$s = k \sin (g\theta - \gamma),$$

we may write it

$$s = k \sin [\theta - \{\gamma - (g-1)\,\theta\}].$$

Let ♈$Nm$ be the ecliptic, $N$ the moon's node when her true longitude is zero, and let $M$ be the position of the moon at time $t$, $m$ her place referred to the ecliptic;

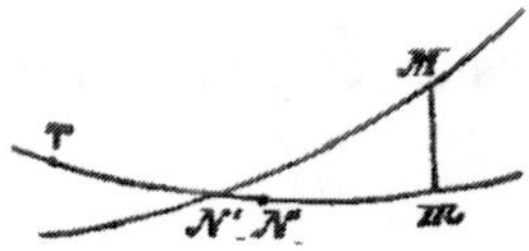


therefore ♈$N = \gamma$, ♈$m = \theta$, $\tan Mm = s$.

Take $NN' = (g-1)\,\theta$ in a retrograde direction; and join $MN'$ by an arc of great circle;

then $\tan Mm = \tan MN'm \sin N'm$,

or $s = \tan MN'm \sin [\theta - \{\gamma - (g-1)\,\theta\}]$;

which, compared with the value of $s$ given above, shews that $MN'm = \tan^{-1} k$ is constant, and therefore the term $k \sin (g\theta - \gamma)$ indicates that the moon moves in an orbit inclined at an angle $\tan^{-1} k$ to the ecliptic, and whose node regredes along the ecliptic with the velocity $(g-1)\dfrac{d\theta}{dt}$, or with a mean velocity $(g-1)\,p$.

86. Hence the period of a revolution of the nodes

$$= \frac{2\pi}{(g-1)\,p} = \frac{\text{one sidereal month}}{g-1},$$

but, from Art. (51), the value of $g = 1 + \frac{3}{4}m^2$;

therefore period of revolution of nodes $= \dfrac{\text{one sidereal month}}{\frac{3}{4}m^2}$

$= 6511$ days, nearly.

This will, for the same reason as in the case of the apse, Art. (75), be modified when we carry the approximation to a higher degree; this value of $g$ is, however, much more accurate than the corresponding value of $c$, for the third term of $g$ is small; the value to the third order being (see Appendix, Art. 102)

$$g = 1 + \tfrac{3}{4}m^2 - \tfrac{9}{32}m^3,$$

and the period of revolution of the nodes $= \dfrac{\text{one sidereal month}}{\frac{3}{4}m^2(1 - \frac{3}{8}m)}$

$= 6705$ days, nearly.

This is not far from the accurate value as given by observation, and when the approximation to the value of $g$ is carried to a higher order, the agreement is nearly perfect.

The true value is 6793·39 days, that is about 18 yrs. 7 mo.

### *Evection in Latitude.*

87. *To explain the variation of the inclination and the irregularity in the motion of the node expressed by the term*

$$+ \tfrac{3}{8}mk \sin\{(2 - 2m - g)\theta - 2\beta + \gamma\}.$$

This term, as a correction on the preceding, is analogous to the evection as a correction on the elliptic inequality.

Taking the two terms together,

$$s = k \sin(g\theta - \gamma) + \tfrac{3}{8}mk \sin\{(2 - 2m - g)\theta - 2\beta + \gamma\}.$$

Let ☽ = longitude of moon $= \theta$,

☉ =............... sun $= m\theta + \beta$,

☊ =............... node $= \gamma - (g - 1)\theta$;

therefore $s = k \sin(☽ - ☊) + \tfrac{3}{8}mk \sin\{☽ - ☊ - 2(☉ - ☊)\}$.

Now these two terms may be combined into one,

$$s = K \sin(☽ - ☊ - \Delta),$$

if $K \cos\Delta = k + \tfrac{3}{8}mk \cos 2(☉ - ☊)$,

$K \sin\Delta = \tfrac{3}{8}mk \sin 2(☉ - ☊)$,

whence $\Delta$ and $K$ may be found; and, approximately,

$$K = k\{1 + \tfrac{3}{8}m \cos 2(\odot - \Omega)\},$$

$$\Delta = \tfrac{3}{8}m \sin 2(\odot - \Omega),$$

but the equation

$$s = K \sin(\text{☽} - \Omega - \Delta)$$

represents motion in an orbit inclined at an angle $\tan^{-1}K$ to the ecliptic, and the longitude of whose node is $\Omega + \Delta$.

This term has therefore the following effects:

1st. The inclination of the moon's orbit is variable, its tangent increases by $\tfrac{3}{8}mk$ when the nodes are in syzygies, relatively to the sun, and decreases by the same quantity when they are in quadrature; the general expression for the increase being $\tfrac{3}{8}mk \cos 2(\odot - \Omega)$.

2nd. The longitude of the node, calculated on supposition of a uniform regression, is increased by $\Delta = \tfrac{3}{8}m \sin 2(\odot - \Omega)$, so that the node is before its mean place while in the first or third quadrant in front of the sun, and behind it in the second and fourth. *Principia*, book III., props. 33 and 35.

The cycle of these changes will be completed in the period of half a revolution of the sun with respect to the node, that is, in 173·21 days, not quite half-a-year.

88. The tangent of the latitude has here been obtained; if we wish to have the latitude itself it will be given by the formula

$$\text{latitude} = s - \tfrac{1}{3}s^3 + \tfrac{1}{5}s^5 - \&\text{c}.$$

which, to the degree of approximation adopted, will clearly be the same as $s$.

## *Reduction.*

89. We may now consider the term which we had neglected (Art. 83) in the expression for the longitude, namely,

$$-\tfrac{1}{4}k^2 \sin 2(gpt - \gamma).$$

Let $N$ be the position of the node when the moon's longitude is $\theta$, $M$ the place of the moon, $m$ the place referred to the ecliptic.

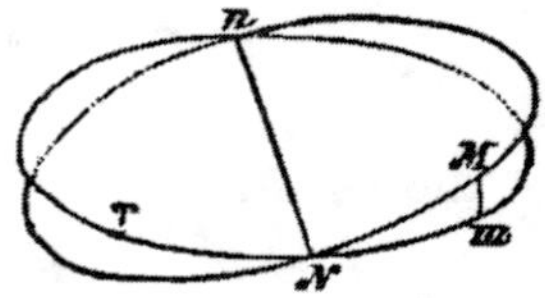

Therefore $\Upsilon m = \theta$,

$$\Upsilon N = \gamma - (g-1)\theta,$$

$$Nm = g\theta - \gamma,$$

$$\tan N = k.$$

The right-angled spherical triangle $NMm$ gives

$$\cos N = \frac{\tan Nm}{\tan NM};$$

therefore
$$\frac{1-\cos N}{1+\cos N} = \frac{\tan NM - \tan Nm}{\tan NM + \tan Nm},$$

or
$$\tan^2 \frac{N}{2} = \frac{\sin(NM - Nm)}{\sin(NM + Nm)},$$

or, since both $N$ and $NM - Nm$ are small,

$$\frac{\tan^2 N}{4} = \frac{NM - Nm}{\sin 2Nm} \text{ approximately};$$

therefore $NM - Nm = \frac{1}{4}k^2 \sin 2(g\theta - \gamma) = \frac{1}{4}k^2 \sin 2(gpt - \gamma)$, nearly.

Hence this term, which is called the *reduction*, is approximately the difference between the longitude in the orbit and the longitude in the ecliptic.

## RADIUS VECTOR.

90. *To explain the physical meaning of the terms in the value of u.*

We shall, for the explanation, make use of the formula which gives the value of $u$ in terms of the true longitude, Art. (52).

Neglecting the periodical terms, we have for the mean value

$$u = a(1 - \tfrac{3}{4}k^2 - \tfrac{1}{2}m^2).$$

The term $-\tfrac{1}{2}m^2$, which is a consequence of the disturbing effect of the sun, shews that the mean value of the moon's radius vector, and therefore the orbit itself, is larger than if there were no disturbance.

*Elliptic Inequality.*

91. *To explain the effect of the term of the first order,*

$$u = a\{1 + e\cos(c\theta - \alpha)\},$$
$$= a[1 + e\cos\overline{\theta - \{\alpha + (1-c)\theta\}}].$$

This is the *elliptic inequality*, and indicates motion in an ellipse whose eccentricity is $e$ and longitude of the apse $\alpha + (1-c)\theta$; and the same conclusion is drawn with respect to the motion of the apse as in Art. (73).

*Evection.*

92. *To explain the physical meaning of the term*

$$\tfrac{15}{8}mea\cos\{(2 - 2m - c)\theta - 2\beta + \alpha\}.$$

This, as in the case of the corresponding term in the longitude, is best considered in connection with the elliptic inequality, and exactly the same results will follow.

Thus, calling ☽, ☉, and $\alpha'$ the true longitudes of the moon, sun, and apse (the last calculated on supposition of uniform motion), these two terms may be written,

$$u = a[1 + e\cos(☽ - \alpha') - \tfrac{15}{8}me\cos\{☽ - \alpha' + 2(\alpha' - ☉)\}]$$
$$= a[1 + E\cos(☽ - \alpha' + \delta)];$$

where
$$E\cos\delta = e + \tfrac{15}{8}me\cos 2(\alpha' - ☉),$$
$$E\sin\delta = \phantom{e +{}} \tfrac{15}{8}me\sin 2(\alpha' - ☉).$$

These are identical with the equations of Art. (77).

### *Variation.*

93. *To explain the effect of the term* $m^2 a \cos\{(2-2m)\theta - 2\beta\}$,

$$u = a\left[1 + m^2 \cos\{(2-2m)\theta - 2\beta\}\right]$$
$$= a\left[1 + m^2 \cos 2\,(☽ - ⊙)\right].$$

As far as this term is concerned, the moon's orbit would be an oval having its longest diameter in quadratures and least in syzygies. *Principia*, lib. I. prop. 66, cor. 4.

The ratio of the axes of the oval orbit will be

$$\frac{1+m^2}{1-m^2} = \frac{90}{89} \text{ nearly, } m \text{ being } \cdot 0748.$$

See *Principia*, lib. III. prop. 28.

### *Reduction.*

94. The last important periodical term in the value of $u$ is

$$-\frac{ak^2}{4}\cos 2\,(g\theta - \gamma).$$

This term expresses approximately the variation in the difference between the values of $u$ in the orbit and in the ecliptic.

For if $u_1$ be the reciprocal of the value of the radius vector in the orbit,

$$u_1 = u \cos(\text{latitude}),$$
$$= \frac{u}{\sqrt{(1+s^2)}} = u\,(1 - \tfrac{1}{2}s^2), \text{ nearly;}$$

therefore
$$u - u_1 = \tfrac{1}{2}us^2 = \tfrac{1}{2}ak^2 \sin^2(g\theta - \gamma)$$
$$= \tfrac{1}{4}ak^2 - \tfrac{1}{4}ak^2 \cos 2\,(g\theta - \gamma)$$
$$= \text{const.} - \tfrac{1}{4}ak^2 \cos 2\,(g\theta - \gamma).$$

Taking this result in connexion with that of Art. (89) we see that the *reduction*, so far as periodical terms to the second order are concerned, is simply a geometrical consequence of the inclination of the orbit; and that, if we measured the longitude and the radius vector along the orbit instead of taking their projection on the plane of the ecliptic, these periodical terms would not appear.

95. There are no other terms of the second order in the value of $u$. The *annual equation*, which, in the longitude, is of the second order, is only of the third order in the radius vector.

### *Periodic time of the Moon.*

96. We have seen, Art. (72), that the periodic time of the moon is greater than if there were no disturbing force; but this refers to the *mean* periodic time established on an interval of a great number of years, so that the circular functions in the expression are then extremely small compared with the quantity $pt$ which has uniformly increased.

When, however, we consider only a few revolutions, these terms may not all be neglected. The *elliptic inequality* and the *evection* go through their values in about a month, the *variation* and *reduction* in about half-a-month; their effects, therefore, on the length of the period can scarcely be considered, as they will increase one portion and then decrease another of the *same* month.

But the *annual equation* takes one year to go through its cycle, and, during this time, the moon has described thirteen revolutions; hence, fluctuations may, and, as we shall now shew, do take place in the lengths of the sidereal months during the year.

We have, considering only the annual equation, Art. (82),

$$pt = \theta + 3me' \sin(m\theta + \beta - \zeta).$$

Let $T$ be the length of the period; then, when $\theta$ is increased by $2\pi$, $t$ becomes $t+T$;

therefore $p(t+T) = 2\pi + \theta + 3me' \sin(2m\pi + m\theta + \beta - \zeta)$,

whence $$pT = 2\pi + 6me' \sin m\pi \cos(m\pi + m\theta + \beta - \zeta);$$

therefore $$T = \text{mean period} + \frac{6me'}{p} \sin m\pi \cos(\odot - \zeta),$$

where $\odot = m\theta + \beta + m\pi$ = sun's longitude at the beginning of the month $+ m\pi$

= sun's longitude at the middle of the month.

Hence $T$ will be longest when $\odot - \zeta = 0$,

and shortest when $\odot - \zeta = \pi$;

or $T$ will be longest when the sun at the middle of the month is in perigee, and shortest when in apogee; but, at present, the sun is in perigee about the 1st of January, and in apogee about the 1st of July; therefore, owing to *annual equation*, the winter months will be longer than the summer months, the difference between a sidereal month in January and July, from this cause, being about 20 minutes.

97. All the inequalities or equations, which our expressions contain, have thus received a physical interpretation. They were the only ones known before Newton had established his theory, but the necessity for such corrections was fully recognized, and the values of the coefficients had already been pretty accurately determined; still, with the exception of the reduction, which is geometrically necessary, they were corrections empirically made, and it was scarcely to be expected that any but the larger inequalities, viz. those of the first and second orders which we have here discussed, could be detected by observation: we find, however, that three others, have, since Newton's time, been indicated by observation before theory had explained their cause. These are—

the *secular acceleration*, discovered by Halley; an inequality, found by Mayer, in the longitude of the moon, and of which the longitude of the ascending node is the argument; and finally an inequality discovered by Bürg, which has only of late years obtained a solution. For a further account of these, as also of some other inequalities which theory has made known, see Appendix, Arts. (108) to (112).

# CHAPTER VII.

## APPENDIX.

In this chapter will be found collected a few propositions intimately connected with the results or the processes of the Lunar Theory as explained in the previous pages. Reference has been made to some of them in the course of the work, and the interest and importance of the others are sufficient to justify their introduction here.

98. *The moon is retained in her orbit by the force of gravity, that is, by the same force which acts on bodies at the surface of the earth.*

The proof of this is merely a numerical verification; the data required from observation are,

the space fallen through from rest in 1″ by bodies at the earth's surface ................................= 16·1 feet,

the radius of the earth ............................= 4000 miles,

the periodic time of the moon ..................= $27\frac{1}{3}$ days,

the distance of the moon from the earth's centre=60×4000 miles.

The force of the earth's attraction $\propto \dfrac{1}{(\text{dist.})^2}$. Therefore, the space fallen through in 1″ at distance of moon by a body moving from rest under the earth's action $= \dfrac{16{\cdot}1}{60^2}$ feet

$= {\cdot}00447$ feet.

But the moon in one second describes an angle $\frac{2\pi}{27\frac{1}{3}.24.60^2}=\omega$, during which the approach to the earth

$$= 60 \times 4000 \times 5280 \text{ (vers. } \omega) \text{ feet}$$

$$= \frac{60 \times 4000 \times 5280 \times 2\pi^2}{(27\frac{1}{3})^2.(24)^2.(60)^4} \text{ feet}$$

$$= \cdot 00448 \text{ feet.}$$

Therefore, the space through which the moon is deflected in one second from her straight path, is just the quantity through which she would fall towards the earth, supposing her to be subject to the earth's attraction; and we may, therefore, conclude that she is retained in her orbit by the force of gravity.

When first Newton, in 1666, attempted to verify this result, he found a difference between the two values equal to one-sixth of the less: the reason of his failure was the incorrect measures of the earth, which he made use of in his computation; and it was not till about 16 years later that he was led to the true result, by using the more correct value of the earth's radius obtained by Picart.

*Principia*, lib. III., prop. 4.

99. *The moon's orbit is everywhere concave to the sun.*

Let $S$, $E$, and $M$ be the centres of the sun, earth, and moon. We ought first to apply to each body forces equal and opposite to those which act on

the sun in order to bring him to rest. These forces are, however, so small that we may neglect them, and we shall consider the moon to be moving about the sun fixed, and to be disturbed by the earth alone.

The forces on $M$ are, therefore, $\dfrac{m'}{SM^2}$ .............. in $MS$,

and $\dfrac{E}{EM^2}$ ..............in $ME$.

This last must be resolved into two, one in $MS$, the other perpendicular to it.

Therefore, the whole central force on the moon in $MS$

$$= \frac{m'}{SM^2} + \frac{E}{EM^2} \cos M,$$

and the proposition will be proved if we shew that this force is always positive. Now, Art. (25),

$$\text{periodic time of sun} = \frac{2\pi . SG^{\frac{3}{2}}}{m'^{\frac{1}{2}}} = \frac{2\pi . SM^{\frac{3}{2}}}{m'^{\frac{1}{2}}} \text{ nearly,}$$

$$\text{and .................... moon} = \frac{2\pi . EM^{\frac{3}{2}}}{(E+M)^{\frac{1}{2}}};$$

$$\text{therefore} \quad \frac{m'}{SM^3} = \tfrac{1}{169} \frac{E+M}{EM^3} \text{ nearly};$$

$$\text{therefore} \quad \frac{m'}{SM^3} > \tfrac{1}{169} \frac{E}{EM^3},$$

$$\frac{m'}{SM^2} > \tfrac{1}{169} \frac{SM}{EM} \cdot \frac{E}{EM^2} > \tfrac{400}{169} \frac{E}{EM^2};$$

$$\text{therefore} \quad \frac{m'}{SM^2} - \frac{E}{EM^2} \text{ is positive:}$$

but this is the value of the central force corresponding to $\cos M = -1$, and is therefore its least value. Hence the central force always tends to the sun, and the path is always concave.

At new moon the force with which the moon tends to the sun is, therefore, greater than that with which she tends to the earth: the earth being itself in motion in the same direction, and, at that instant, with greater velocity, will easily explain how, notwithstanding this, the moon still revolves about it.*

*Central and Tangential Disturbing Forces.*

100. We have hitherto considered the effects of the central and tangential disturbing forces in combination; but it will be interesting to determine to which of them the several inequalities principally owe their existence.

(1) *To determine the effect of the central disturbing force.*

Make $T=0$;

therefore $$\frac{d^2u}{d\theta^2}+u-\frac{P}{h^2u^2}=0,$$

or substituting for $\frac{P}{h^2u^2}$ from Art. (52),

$$\frac{d^2u}{d\theta^2}+u=a\left\{\begin{array}{l}1-\frac{3}{4}k^2-\frac{1}{2}m^2+\frac{3}{2}m^2e\cos c\theta_1-\frac{3}{2}m^2\cos(2-2m)\theta_1\\ \qquad+\frac{9}{4}m^2e\cos(2-2m-c)\theta_1\\ \qquad+\frac{3}{4}k^2\cos 2g\theta_1;\end{array}\right.$$

therefore,

$$u=a\left\{\begin{array}{l}1-\frac{3}{4}k^2-\frac{1}{2}m^2+e\cos(c\theta-\alpha)+\frac{1}{2}m^2\cos\{(2-2m)\theta-2\beta\}\\ \qquad+\frac{9}{16}me\cos\{(2-2m-c)\theta-2\beta+\alpha\}\\ \qquad-\frac{1}{4}k^2\cos 2(g\theta-\gamma).\end{array}\right.$$

If we compare this with the value of $u$ found Art. (52), we see that the elliptic inequality and the reduction are due to the central or radial force, as also one-half of the variation and three-tenths of the evection.

It would perhaps be proper to separate the absolute central force from the central *disturbing* force; the terms due to the latter are those which contain $m$; therefore, the elliptic

* See the Author's *Astronomy*, p. 233.

inequality and the reduction are the effects of the former, except that in the elliptic inequality the introduction of $c$, or the motion of the apse, is due to the disturbing force.

(2) *To determine the effect of the tangential disturbing force.*

Let the central *disturbing* force be zero;

then $\dfrac{P}{h^2u^2} = \dfrac{\mu}{h^2}(1 - \frac{3}{2}s^2) = a$, neglecting the inclination, and, substituting for $\dfrac{T}{h^2u^3} \cdot \dfrac{du}{d\theta}$ and for $2\left(\dfrac{d^2u}{d\theta^2} + u\right)\displaystyle\int \dfrac{T}{h^2u^3}\,d\theta$ their values from Art. (52), the equation in $u$ becomes

$$\frac{d^2u}{d\theta^2} + u = a\left\{1 - \tfrac{3}{2}m^2 \cos(2-2m)\,\theta_1 + \tfrac{21}{4}m^2 e \cos(2-2m-c)\,\theta_1\right\};$$

$$\text{whence } u = a\begin{cases} 1 + e\cos(c\theta - \alpha) + \frac{1}{2}m^2 \cos\{(2-2m)\,\theta - 2\beta\} \\ \quad + \frac{21}{16}me \cos\{(2-2m-c)\,\theta - 2\beta + \alpha\}. \end{cases}$$

We have here the remaining half of the variation and the remaining seven-tenths of the evection as the effects of the tangential disturbance. Also $c = 1$, or, to the second order, the tangential force has no effect on the motion of the apse.

101. The separate effects of the central and tangential forces in producing the inequalities in the longitude may be traced in a similar manner. We shall leave this as an exercise for the student. The annual equation will be found to be due to the central force. The other inequalities will be divided as in the radius vector; except as regards the variation, of which four-elevenths are due to the central and the remaining seven-elevenths to the tangential force.

*To calculate the value of $g$ to the third order.*

102. We must here make use of the results which the approximations to the second order have furnished; but as the value of $g$ is determined by that term of the differential equation whose argument is $g\theta - \gamma$, we need only consider

those terms which by their combinations will lead to it without rising to a higher order than the fourth.

From Arts. (51), (31), we obtain

$$s = k \sin g\theta_1 + \tfrac{3}{8} mk \sin (2 - 2m - g)\, \theta_1,$$

$$\frac{Ps - S}{h^2 u^3} = -\tfrac{3}{2} m^2 s \{1 + \cos (2 - 2m)\, \theta_1\},$$

$$\frac{T}{h^2 u^3} = -\tfrac{3}{2} m^2 \sin (2 - 2m)\, \theta_1.$$

From these we obtain

$$\frac{Ps - S}{h^2 u^3} = \ldots - \tfrac{3}{2} m^2 k \left(1 - \tfrac{3}{16} m\right) \sin g\theta_1$$

$$\frac{T}{h^2 u^3} \cdot \frac{ds}{d\theta} = \ldots - \tfrac{3}{2} m^2 k \sin (2 - 2m)\, \theta_1 \{\cos g\theta_1 + \tfrac{3}{8} m \cos (2 - 2m - g)\, \theta_1\}$$

$$= \ldots - \tfrac{9}{32} m^3 k \sin g\theta_1,$$

$$\frac{d^2 s}{d\theta^2} + s = 0, \text{ to the second order;}$$

$$\text{therefore } 2 \left(\frac{d^2 s}{d\theta^2} + s\right) \int \frac{T}{h^2 u^3}\, d\theta = 0, \text{ to the fourth order.}$$

Therefore the equation in $s$ becomes, so far as these terms are concerned,

$$\frac{d^2 s}{d\theta^2} + s = \left(-\tfrac{3}{2} m^2 k + \tfrac{9}{32} m^3 k + \tfrac{9}{32} m^3 k\right) \sin g\theta_1;$$

assume $$s = k \sin g\theta_1,$$

therefore $$k(1 - g^2) = -\tfrac{3}{2} m^2 k + \tfrac{9}{16} m^3 k,$$

$$g = 1 + \tfrac{3}{4} m^2 - \tfrac{9}{32} m^3.$$

*To find the value of c to the third order.*

103. Proceeding in a similar manner from the equations, Arts. (52), (31), we get

$$u = a \{1 + e \cos c\theta_1 + m^2 \cos (2 - 2m)\, \theta_1 + \tfrac{15}{8} me \cos (2 - 2m - c)\, \theta_1\},$$

N

$$\frac{P}{h^2u^2} = a - \tfrac{1}{2}m^2\frac{a^4}{u^3}\{1 + 3\cos(2-2m)\theta_1\},$$

$$\frac{T}{h^2u^3} = \quad -\tfrac{3}{2}m^2\frac{a^4}{u^4}\sin(2-2m)\theta_1,$$

we obtain

$$\frac{P}{h^2u^2} = a - \tfrac{1}{2}m^2a\{1 - 3e\cos c\theta_1 \ldots - \tfrac{45}{8}me\cos(2-2m-c)\theta_1\}$$
$$\times\{1 + 3\cos(2-2m)\theta_1\}$$
$$= a + \tfrac{1}{2}m^2a\{3e + \tfrac{135}{16}me\}\cos c\theta_1,$$

$$\frac{T}{h^2u^3} = -\tfrac{3}{2}m^2\{1 - 4e\cos c\theta_1 - 4m^2\cos(2-2m)\theta_1 - \tfrac{15}{2}me\cos(2-2m-c)\theta_1\}$$
$$\times\sin(2-2m)\theta_1$$
$$= -\tfrac{3}{2}m^2\{\sin(2-2m)\theta_1 - 2e\sin(2-2m-c)\theta_1 - \tfrac{15}{4}me\sin c\theta_1\},$$

$$\frac{du}{d\theta} = -a\{e\sin c\theta_1 + 2m^2\sin(2-2m)\theta_1 + \tfrac{15}{8}me\sin(2-2m-c)\theta_1\};$$

therefore $$\frac{T}{h^2u^3}\cdot\frac{du}{d\theta} = \ldots\tfrac{45}{32}m^3ae\cos c\theta_1,$$

the other term is of the fifth order.

$$\int\frac{T}{h^2u^3}d\theta = +\tfrac{3}{4}m^2\cos(2-2m)\theta_1 - 3m^2e\cos(2-2m-c)\theta_1 - \tfrac{45}{8}m^3e\cos c\theta_1,$$

$$\left(\frac{d^2u}{d\theta^2} + u\right) = a\{1 \ldots - 3m^2\cos(2-2m)\theta_1 + \tfrac{15}{2}m^2e\cos(2-2m-c)\theta_1;$$

therefore $$2\left(\frac{d^2u}{d\theta^2} + u\right)\int\frac{T}{h^2u^3}d\theta = -\tfrac{45}{4}m^3ae\cos c\theta_1,$$

and the equation in $u$ becomes

$$\frac{d^2u}{d\theta^2} + u = a\{1 + (\tfrac{3}{2}m^2e + \tfrac{135}{32}m^3e - \tfrac{45}{32}m^3e + \tfrac{45}{4}m^3e)\cos c\theta_1 + \ldots\}$$
$$= a\{1 + (\tfrac{3}{2}m^2e + \tfrac{225}{16}m^3e)\cos c\theta_1 + \ldots\}.$$

Assume $$u = a(1 + e\cos c\theta_1 + \ldots);$$

therefore $$ae(1 - c^2) = (\tfrac{3}{2}m^2e + \tfrac{225}{16}m^3e)a,$$

whence $$c = 1 - \tfrac{3}{4}m^2 - \tfrac{225}{32}m^3.$$

104. Hence, to the third order of approximation,

$$\frac{\text{mean motion of apse}}{\text{mean motion of node}} = \frac{1-c}{g-1} = \frac{\frac{3}{4}m^2 + \frac{225}{32}m^3}{\frac{3}{4}m^2 - \frac{9}{32}m^3} = \frac{8+75m}{8-3m},$$

and since $m = \frac{1}{13}$ nearly, we see that the moon's apse progredes nearly twice as fast as the node regredes.

In the case of one of Jupiter's satellites, the periodic time round Jupiter is only a few of our days, and the periodic time of Jupiter round the sun is 12 of our years, therefore $m$, the ratio of these periods, is very small.

Hence, the apse of one of Jupiter's satellites progredes along Jupiter's ecliptic, with pretty nearly the same velocity as the node regredes, assuming these motions to be due to the sun's disturbing force; they are, however, principally due to the oblateness of the planet.

## *Parallactic Inequality.*

105. In carrying on the approximations to a higher order, it is found that the expressions for the forces contain terms whose argument is the moon's *elongation*, *i.e.* the difference of longitude of the sun and moon. These terms appear in $P$ and in $T$ of the fourth order, and since $1-m$, the coefficient of $\theta$, is near unity, the terms will become of the third order, and therefore of considerable importance in the values of $u$ and of $t$. We shall work out these expressions from the earliest steps; and we select these terms on account of the peculiar use which has been made of them to determine the sun's parallax, whence they have received the name of parallactic inequality.

Let us go back to the expressions for the forces Arts. (23), (24). In the first place we may remark that the terms we are about to compute must be independent of $s$; because any term into the composition of which $s$ enters will necessarily have $g$ in the coefficient of the argument. We may therefore at the outset suppose the orbit to coincide with the ecliptic. We have then

$$SM^2 = r'^2 - 2r'MG\cos(\theta-\theta') + MG^2,$$

$$\frac{1}{SM^3} = \frac{1}{r'^3}\left[1+\frac{3MG}{r'}\cos(\theta-\theta') - \frac{3}{2}\frac{MG^2}{r'^2} + \frac{15}{2}\frac{MG^2}{r'^2}\cos^2(\theta-\theta')\right]$$

$$= \frac{1}{r'^3}\left[1+\frac{3MG}{r'}\cos(\theta-\theta') + \frac{MG^2}{r'^2}\left\{\tfrac{9}{4}+\tfrac{15}{4}\cos 2(\theta-\theta')\right\}\right],$$

so $$\frac{1}{SE^3} = \frac{1}{r'^3}\left[1-\frac{3EG}{r'}\cos(\theta-\theta') + \frac{EG^2}{r'^2}\left\{\tfrac{9}{4}+\tfrac{15}{4}\cos 2(\theta-\theta')\right\}\right].$$

We must now substitute these values in

$$P = \frac{E+M}{ME^2} + m'\left(\frac{MG}{SM^3}+\frac{EG}{SE^3}\right) - m'r'\left(\frac{1}{SM^3}-\frac{1}{SE^3}\right)\cos(\theta-\theta'),$$

$$T = \qquad - m'r'\left(\frac{1}{SM^3}-\frac{1}{SE^3}\right)\sin(\theta-\theta'),$$

and pick out the terms which have for argument $(\theta-\theta')$.

In $m'\left(\frac{MG}{SM^3}+\frac{EG}{SE^3}\right)$ there will be a term

$$3m'\frac{MG^2-EG^2}{r'^4}\cos(\theta-\theta') = \frac{3m'r^2}{r'^4}\,\frac{E-M}{E+M}\cos(\theta-\theta').$$

In $m'r'\left(\frac{1}{SM^3}-\frac{1}{SE^3}\right)\cos(\theta-\theta')$ we shall have the terms

$$m'\frac{(MG^2-EG^2)}{r'^4}\left\{\tfrac{9}{4}\cos(\theta-\theta') + \tfrac{15}{4}\cos(\theta-\theta')\cos 2(\theta-\theta')\right\},$$

which produce

$$m'\frac{\{MG^2-EG^2\}}{r'^4}\left(\tfrac{9}{4}+\tfrac{15}{8}\right)\cos(\theta-\theta') = \tfrac{33}{8}\frac{m'r^2}{r'^4}\,\frac{E-M}{E+M}\cos(\theta-\theta'),$$

and finally $m'r'\left(\frac{1}{SM^3}-\frac{1}{SE^3}\right)\sin(\theta-\theta')$ will produce

$$m'\frac{MG^2-EG^2}{r'^4}\left(\tfrac{9}{4}-\tfrac{15}{8}\right)\sin(\theta-\theta') = \tfrac{3}{8}\frac{m'r^2}{r'^4}\,\frac{E-M}{E+M}\sin(\theta-\theta').$$

Therefore the new terms will be

$$\text{in } \frac{P}{h^2u^2} \ldots - \tfrac{9}{8}\frac{m'u'^4}{h^2u^4}\frac{E-M}{E+M}\cos(\theta-\theta') = -\tfrac{9}{8}m^2 a' \frac{E-M}{E+M}\cos(\theta-\theta'),$$

$$\text{in } \frac{T}{h^2u^3} \ldots - \tfrac{3}{8}\frac{m'u'^4}{h^2u^5}\frac{E-M}{E+M}\sin(\theta-\theta') = -\tfrac{3}{8}m^2 \frac{a'}{a}\frac{E-M}{E+M}\sin(\theta-\theta').$$

Since $\frac{a'}{a}$ ($=\frac{1}{400}$ nearly) is of the second order, these terms are of the fourth order.

106. On the principle of the superposition of small disturbances, we may compute the effects of these terms by themselves as disturbers of the elliptic path.

$$\frac{P}{h^2u^2} = a\left[1 + \ldots\ldots - \tfrac{9}{8}m^2\frac{E-M}{E+M}\frac{a'}{a}\cos\{(1-m)\theta-\beta\}\right],$$

$$\frac{T}{h^2u^3} = \ldots\ldots\ldots\ldots\ldots\ldots - \tfrac{3}{8}m^2\frac{E-M}{E+M}\frac{a'}{a}\sin\{(1-m)\theta-\beta\},$$

$$u = a\{1 + e\cos(c\theta-\alpha) + \ldots\ldots\},$$

$$\frac{du}{d\theta} = -ae\sin(c\theta-\alpha)\ldots\ldots;$$

therefore

$$\frac{T}{h^2u^3}\frac{du}{d\theta} = 0, \text{ to the fourth order,}$$

$$\frac{d^2u}{d\theta^2} + u = a, \text{ to the second order,}$$

$$\int\frac{T}{h^2u^3}d\theta = \ldots\ldots + \tfrac{3}{8}m^2\frac{E-M}{E+M}\frac{a'}{a}\cos\{(1-m)\theta-\beta\}.$$

Substituting in the differential equation for $u$, we get

$$\frac{d^2u}{d\theta^2} + u = a\left[1 + \ldots\ldots - \tfrac{15}{8}m^2\frac{E-M}{E+M}\frac{a'}{a}\cos\{(1-m)\theta-\beta\} + \ldots\right].$$

Assume $u = a\left[1 + \ldots\ldots\ldots + A\frac{E-M}{E+M}\frac{a'}{a}\cos\{(1-m)\theta-\beta\} + \ldots\right]$;

therefore $$A = \frac{-\frac{15}{8}m^2}{1-(1-m)^2} = -\tfrac{15}{16}m\,;$$

therefore $$u = a\left[1 + \ldots\ldots - \tfrac{15}{16}m\,\frac{E-M}{E+M}\frac{a'}{a}\cos\{(1-m)\,\theta - \beta\} + \ldots\right].$$

107. The corresponding term in the value of $\theta$ will also be of the third order,

$$\frac{dt}{d\theta} = \frac{1}{hu^2}\left(1 - \int \frac{T}{h^2u^3}\,d\theta\right),$$

$$\frac{1}{u^2} = \frac{1}{a^2}\left[1 + \ldots\ldots + \tfrac{15}{8}m\,\frac{E-M}{E+M}\frac{a'}{a}\cos\{(1-m)\,\theta - \beta\} + \ldots\ldots\right],$$

$\int \frac{T}{h^2u^3}\,d\theta$ is of the fourth order and will not rise in $t$;

therefore

$$\frac{dt}{d\theta} = \frac{1}{p}\left[1 + \ldots\ldots + \tfrac{15}{8}m\,\frac{E-M}{E+M}\frac{a'}{a}\cos\{(1-m)\,\theta - \beta\} + \ldots\right],$$

$$t = \frac{1}{p}\left[\theta + \ldots\ldots + \tfrac{15}{8}m\,\frac{E-M}{E+M}\frac{a'}{a}\sin\{(1-m)\,\theta - \beta\} + \ldots\right],$$

and $$\theta = pt + \ldots\ldots - \tfrac{15}{8}m\,\frac{E-M}{E+M}\frac{a'}{a}\sin\{(1-m)\,pt - \beta\} + \ldots.$$

Mayer was the first who applied this term to the determination of the sun's parallax, by comparing the analytical expression of the coefficient with its value as deduced from observation. The values of $m$ and of $\frac{E}{M}$, and therefore of $\frac{E-M}{E+M}$ being pretty accurately known, $\frac{a'}{a}$ will be determined, that is, the ratio of the sun's parallax to that of the moon; but the moon's parallax is well known; therefore, also, that of the sun can be calculated. The value so obtained by Laplace was 8·6″; but Hansen, by the discussion of a great number of modern observations, has found 8·92″, a value which agrees very closely with recent determinations by other methods.

### *Secular Acceleration.*

108. Halley, about 1693, found, by the comparison of ancient and modern eclipses, that the moon's mean revolution is now performed in a shorter time than at the epoch of the recorded Chaldean and Babylonian eclipses. This phenomenon, called the *secular acceleration of the moon's mean motion*, has not yet been fully accounted for. For a long time it was altogether unexplained, but in 1787 Laplace gave what seemed to be a satisfactory and complete account of it.

The value of $p$, Art. (54), on which the length of the mean period depends, is found, when the approximation is carried to a higher order, to contain the quantity $e'$ the excentricity of the earth's orbit. Now, this excentricity is undergoing a slow but continual change from the action of the planets, and therefore $p$, as deduced from observations made in different centuries, will have different values.

The value of $p$ is at present increasing, or the mean motion is being accelerated, and it will continue thus to increase for a period of immense, but not infinite duration; for, as shewn by Lagrange, the actions of the planets on the excentricity of the earth's orbit will be ultimately reversed, $e'$ will cease to diminish and begin to increase, and consequently $p$ will begin to decrease, and the *secular acceleration* will, so far as this cause is concerned, become a *secular retardation.*

It is worthy of remark that the action of the planets on the moon, thus transmitted through the earth's orbit, is more considerable than their direct action.

109. In his investigation, Laplace treated the excentricity $e'$ of the earth's orbit as a constant in the differential equations, and only considered its variation in the final result. This of course greatly simplified the work, but ought to have been looked upon only as a first approximation; yet, strange to say, the result so obtained agreed almost perfectly with that which had been deduced from a comparison of ancient and modern observations. "Malheureusement cette merveilleuse concordance à laquelle Laplace a dû peut-être une partie de l'éclat de sa plus belle découverte, est sensiblement altérée par les approximations suivantes."*

Some years ago Professor Adams took up the question *ab initio* without any limitation, and introduced the variability of $e'$ into the differential equations themselves. This strictly correct process gave a value of the secular variation only about half of Laplace's which had agreed so well with observation.

The variation of excentricity does not therefore account for the whole observed change, and the mean motion of the moon is affected by some other cause or causes at present unknown. Perhaps the remaining acceleration is only apparent, and arises from a gradual retardation of the earth's diurnal motion. The effect would be precisely the same, and such a retardation may be due to the action of the tides. Whatever be the cause, the explanation of this unexplained phenomenon is a question worthy of the mathematician's most serious efforts.

### *Inequalities depending on the Figure of the Earth.*

110. The earth, not being a perfect sphere, will not attract as if the whole of its mass were collected at its centre: hence, some correction must be introduced to take into account this want of sphericity, and some relation must exist

* Pontécoulant, *Système du Monde*. Supplement, 1860.

between the oblateness and the disturbance it produces. Laplace in examining its effect found that it satisfactorily explained the introduction of a term in the longitude of the moon, which Mayer had discovered by observation, and the argument of which is the true longitude of the moon's ascending node.

By a comparison of the observed and theoretical values of the coefficient of this term, we may determine the oblateness of the earth with as great accuracy as by actual measures on the surface.

111. By pursuing his investigations, with reference to the oblateness, in the expression for the moon's latitude, Laplace found that it would there give rise to a term in which the argument was the true longitude of the moon.

This term, which was uususpected before, will also serve to determine the earth's oblateness, and the agreement with the result of the preceding is almost perfect, giving the compression $\frac{1}{305}$,* which is about a mean between the different values obtained by other methods.

*Perturbations due to Venus.*

112. After the expression for the moon's longitude had been obtained by theory, it was found that there was still a slight deviation between her calculated and observed places, and Bürg, who discovered it by a discussion of the observations of Lahire, Flamsteed, Bradley, and Maskelyne, thought it could be represented by an inequality whose period would be 184 years and coefficient 15″. This was entirely conjectural, and though several attempts were made, it was not accounted for by theory.

* Pontécoulant, *Système du Monde*, vol. IV.

About 1848, Professor Hansen, of Seeberg, in Gotha, having commenced a revision of the Lunar Theory, found two terms, which had hitherto been neglected, due to the action of Venus. One of them is direct and arises from a 'remarkable numerical relation between the anomalistic 'motions of the moon and the sidereal motions of Venus 'and the earth; the other is an indirect effect of an inequality 'of long period in the motions of Venus and the earth, 'which was discovered some years ago by the Astronomer 'Royal.'*

The periods of these two inequalities are extremely long, one being 273 and the other 239 years, and their coefficients are respectively 27·4″ and 23·2″. 'These are considerable 'quantities in comparison with some of the inequalities already 'recognised in the moon's motion, and, when applied, they 'are found to account for the chief, indeed the only 'remaining, empirical portion of the moon's motion in 'longitude of any consequence; so that their discovery may 'be considered as a practical completion of the Lunar Theory, 'at least for the present astronomical age, and as establishing 'the entire dominion of the Newtonian Theory and its 'analytical application over that refractory satellite.'†

### *Motion of the Ecliptic.*

113. We have seen, Art. (14), that our plane of reference is not a fixed plane, but its change of position is so slow that we have been able to neglect it, and it is only when the approximation is carried to a higher order that the necessity arises for taking account of its motion.

---

* Report to the Annual General Meeting of the Royal Astronomical Society, Feb. 11, 1848.

† Address of Sir John Herschel to the Meeting of the Royal Astronomical Society.

It has been found to have an angular velocity, about an axis in its own plane, of 48″ in a century, and the correction thus introduced produces in the latitude of the moon a term

$$-c\omega \cos(\theta - \phi),$$

where $\omega$ is the angular velocity of the ecliptic, $\frac{1}{c}$ the angular velocity with which the ascending node of the moon's orbit recedes from the instantaneous axis about which the ecliptic rotates, $\phi$ the longitude of this axis at time $t$, and $\theta$ the longitude of the moon at the same instant.

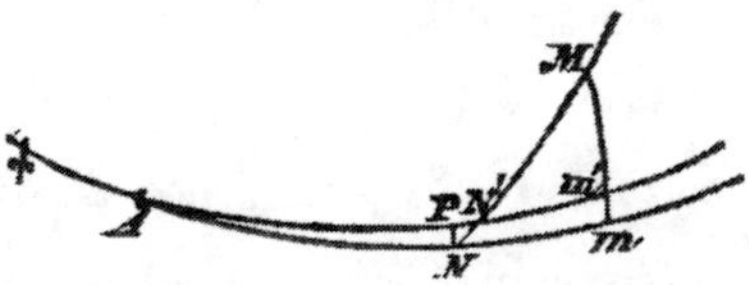


Let $\Upsilon Am$ be the position of the ecliptic at time $t$,

$A$ the point about which it is turning, $\Upsilon A = \phi$,

$MN$ the moon's orbit, $M$ the moon, and $Mm$ a perpendicular to the ecliptic; $\Upsilon m = \theta$; $Mm = \text{lat.} = \beta$.

$i$ the inclination of the orbit, and $N$ the longitude of the node.

Let $APN'm'$ be the ecliptic after a time $\delta t$.

Any point whose longitude is $L$ may be considered as moving perpendicularly to the ecliptic with a velocity

$$\omega \sin(L - \phi).$$

Hence, the point $N$ will move in the direction $PN$ with a velocity $\omega \sin(N - \phi)$. And $N'$ will move along $PN'$ with a velocity $\omega \sin(N - \phi) \cot i$;

therefore $$\frac{dN}{dt} = \omega \sin(N - \phi) \cot i.$$

Again, the point of the ecliptic 90° in advance of $N$ will move towards the moon's orbit with a velocity

$$\omega \sin(90^\circ + N - \phi);$$

therefore $$\frac{di}{dt} = -\omega \cos (N - \phi).$$

Now, $\cot i$, $\omega$, and $\frac{d(N-\phi)}{dt} = -\frac{1}{c}$ may be considered constant in integrating;

therefore $$\delta N = c\omega \cos (N - \phi) \cot i,$$
$$\delta i = c\omega \sin (N - \phi),$$

and if $NM = \psi$, we have

$$\delta\psi = -\frac{\delta N}{\cos i} = -\frac{c\omega}{\sin i} \cos (N - \phi).$$

Now, $\sin \beta = \sin i . \sin \psi$;

$$\therefore \quad \cos \beta . \delta\beta = \cos i . \sin \psi . \delta i + \sin i . \cos \psi . \delta\psi$$
$$= c\omega \{\cos i \sin \psi \sin (N - \phi) - \cos \psi \cos (N - \phi)\},$$

but $\cos i . \sin \psi = \cos \beta \sin (\theta - N)$ and $\cos \psi = \cos \beta \cos (\theta - N)$;

therefore $\delta\beta = -c\omega \cos (\theta - \phi)$.

The discovery of this term is due to Professor Hansen; its coefficient is extremely small, about 1·5″; but, being of a totally different nature from those due to successive approximations, it was thought desirable to examine it, and the above investigation, which was communicated to me by J. C. Adams, Esq.,* will be read with interest on account of its elegance.

With respect to the foregoing investigation, perhaps the following remarks will not be superfluous:—The value of $\frac{dN}{dt}$ is not the actual velocity of $N$, but its velocity relatively to the position of the node as determined when the motion of the ecliptic is neglected; its integral is therefore $\delta N$ the change of longitude due to this motion, and in this integration no constant is added, zero being taken for the mean value. The periodic forms of both $\frac{dN}{dt}$ and $\delta N$ shew that they oscillate about mean values, the time of a complete oscillation being

---

* Now Lowndean Professor of Astronomy in the University of Cambridge.

that reqnired by sin $(N-\phi)$ and cos $(N-\phi)$ to go through their cycle. This relative motion of the node is analogous to that of a particle moving in a straight line under the action of a force varying directly as the distance.

Similar remarks apply to the inclination.

It must also be borne in mind that the result obtained $\delta\beta$ is not the difference between the latitude referred to the actual ecliptic and that referred to a fixed plane, but is the difference between the calculated latitude referred to the actual ecliptic on supposition of its being fixed, and the correct latitude referred to the same actual ecliptic when its motion is taken into account.

We may obtain an approximate value of the coefficient $c\omega$ by substituting for it $\frac{n\alpha}{2\pi}$, where $\alpha$ is the number of seconds through which the ecliptic is deflected in one year $= 0{\cdot}48''$, and $n$ is the number of years in which the node of the moon's orbit makes a complete revolution $= 18{\cdot}6$; for then, $\frac{2\pi}{n}$ is the angle described by the node in one year; therefore $\frac{n\alpha}{2\pi}$ is the ratio of $\omega : \frac{d(N-\phi)}{dt}$, supposing $\phi$ to remain constant, which is nearly the case; therefore

$$c\omega = \frac{n\alpha}{2\pi} = \frac{9{\cdot}3 \times 0{\cdot}48''}{3{\cdot}14} = 1{\cdot}42''.*$$

* This affords the solution of a problem proposed in the Senate-House in the January Examination of 1852. Question 21, Jan. 22.

# CHAPTER VIII.

## HISTORY OF THE LUNAR PROBLEM BEFORE NEWTON.

114. The idea which most probably suggested itself to the minds of those men who first considered the motion of the moon among the stars, was that it described a circle with uniform velocity about the earth as a centre; and this first rough result is represented in our equations by neglecting all small terms and writing $\theta = pt$, $u = a$.

It must, however, have been very soon perceived that the actual motion is far from being so simple, and that the moon moves with very different velocities at different times.

115. The earliest recorded attempts to take into account the irregularites of the moon's motion were made by Hipparchus (140 B.C.). He imagined the moon to move with uniform velocity in a circle, of which the earth occupied, not the centre, but a point nearer to one side. By a similar hypothesis he had accounted for the irregularities in the sun's motion, and his success in this led him to apply it also to the moon.

It is clear that, on this supposition, the moon would seem to move faster when nearest the earth or in perigee, and slower when in apogee, than at any other points of her orbit, and thus an apparent unequal motion would be produced.

Let $BAM$ be a circle, $CA$ a radius, $E$ a point in $AC$ near $C$; $CB$, $ED$ two parallel lines making an angle $\alpha$ with $CA$.

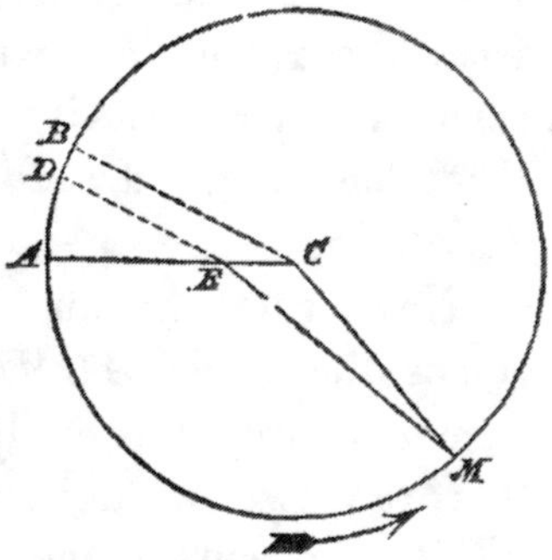


Suppose a body $M$ to describe this circle uniformly with an angular velocity $p$, the time being reckoned from the instant when the body was at $B$, and the longitude as seen from $E$ being reckoned from the line $ED$;

therefore $\quad DEM = \theta, \qquad BCM = pt,$

$$AEM = \theta - \alpha, \quad ACM = pt - \alpha.$$

Now $\frac{EC}{CM}$ is a small fraction, and if we represent it by $e$, we shall have

$$\begin{aligned} \sin M &= \frac{EC}{CM} \sin AEM \\ &= e \sin(\theta - \alpha) \\ &= e \sin(pt + M - \alpha), \end{aligned}$$

$$\tan M = \frac{e \sin(pt - \alpha)}{1 - e \cos(pt - \alpha)};$$

this would give $M$, and then $\theta$ by the formula $\theta = pt + M$.

This was called an *excentric*, and the value of $e$ was called the *excentricity*, which, for the moon, Hipparchus fixed at $\sin 5° \, 1'$.

116. Another method of considering the motion was by means of an *epicycle*, which led to the same result.

A small circle $PM$, with a radius equal to $EC$ of previous figure, has its centre in the circumference of the circle $RPD$ (which has the same radius as that of the excentric), and moves round $E$ with the uniform angular velocity $p$, the body $M$ being carried in the circumference of the smaller circle, the radius $PM$ remaining parallel to itself, or, which is the same thing, revolving from the radius $PE$ with the same angular velocity $p$, so that the angle $EPM$ equals $PEA$.

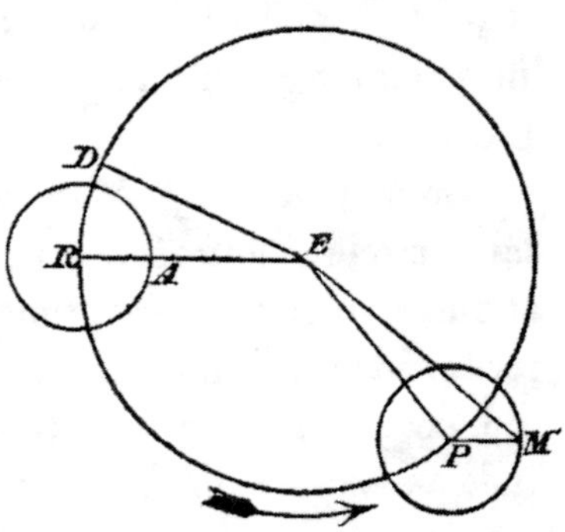


Now, when the angle $AEP$ equals the angle $ACM$ of the former figure, it is easily seen that the two triangles $EPM$, $ECM$ are equal, and therefore the distance $EM$ and the angle $AEM$ will be the same in both, and the two motions will be identical.

**117.** The value of $e$ being small, we find, rejecting $e^2$, &c.,

$$M = e \sin (pt - \alpha),$$

therefore
$$\theta = pt + e \sin (pt - \alpha).$$

If we reject terms of the second order in our expression for the longitude, and make $c = 1$, we get, Art. (55),

$$\theta = pt + 2e \sin (pt - \alpha),$$

which will be identical with the above if we suppose the excentricity of the excentric to be double that of the elliptic orbit.

Ptolemy (A.D. 140) calculated the excentricity of the moon's orbit, and found for it the same value as Hipparchus, viz.

$$\sin 5° \, 1' = \tfrac{1}{12}, \text{ nearly}.$$

The excentricity in the elliptic orbit is, we know, about $\frac{1}{20}$, and $\frac{1}{12}$ and $\frac{1}{20}$ will pretty nearly reconcile the two values of $\theta$ given above. This shews us, that for a few revolutions the moon may be considered as moving in an excentric, and her positions in longitude calculated on this supposition will be correct to the first order.

Her distances from the earth will not however agree; for the ratio of the calculated greatest and least distances would be $\frac{1+\frac{1}{12}}{1-\frac{1}{12}}$ or $\frac{13}{11}$, while that of the true ones would be $\frac{1+\frac{1}{20}}{1-\frac{1}{20}}$ or $\frac{21}{19}$.

It would, therefore, have required two different excentrics to account for the changes in the moon's longitude and in her radius vector. Changes in the latter could not, however, be easily observed with the rude instruments the ancients possessed, and it was very long before this inconsistency was detected.

118. We have said that the moon's longitude, calculated on the hypothesis of an excentric, will be pretty accurate for a *few* revolutions.

The data requisite for this calculation are, the mean angular motion of the moon, the position of the apogee, and the magnitude of the excentricity.

But it was known to Hipparchus and to the astronomers of his time, that the point of the moon's orbit where she seems to move slowest, is constantly changing its position among the stars. Now this point is the apogee of Hipparchus's excentric, and he found that he could very conveniently take account of this further change by supposing the excentric itself to have an angular motion about the earth in the same direction as the moon, so as to make a complete revolution in about nine years, or about 3° in each revolution.*

This motion of the apsidal line follows also from our expression for the longitude, as shewn in Art. (75). It is there, however, connected with an ellipse instead of an

* On the supposition of an epicycle, this motion of the apse could as easily be represented by supposing the radius which connects the moon with the centre of the epicycle to have this uniform angular velocity of about 3° in each revolution, and also in the same direction.

excentric; and though the discovery that the elliptic is the true form of the fundamental orbit was not the next in the order of time after those of Hipparchus, yet, as all the irregularities which were discovered in the intervening seventeen centuries are common both to Hipparchus's excentric and to Kepler's ellipse, it will be as well for us to consider at once this new form of the orbit.

### *Elliptic Form of the Orbit.*

119. We need not dwell on the steps which led to this great and important discovery. Kepler, finding that the predicted places of the planet Mars, as given by the circular theories then in use, did not always agree with the computed ones, sought to reconcile these variances by other combinations of circular orbits, and after a great number of attempts and failures, and eight years of patient investigation, he found it necessary to discard the excentrics and epicycles altogether, and to adopt some new supposition. An ellipse with the sun in the focus was at last his fortunate hypothesis, which was found to give results in accordance with observation; and this form of the orbit was, with equal success, afterwards extended to the moon: but the departures from elliptic motion, due to the disturbing force of the sun, are, in the case of the moon, much greater than the disturbances of the planet Mars by the other planets.

In Keplers's hypothesis, then, the earth is to be considered as occupying the focus of an ellipse, in the perimeter of which the moon is moving, no longer with either uniform linear or angular velocity, but in such a manner that the radius vector sweeps over equal areas in equal times.

This agrees with our investigation of the motion of two bodies, Art. (10).

*Evection.*

120. The hypothesis of an excentric, whose apse line has a progressive motion, as conceived by Hipparchus, served to calculate with considerable accuracy the circumstances or eclipses; and observations of eclipses, requiring no instruments, were then the only ones which could be made with sufficient exactness to test the truth or fallacy of the supposition.

Ptolemy (A.D. 140) having constructed an instrument, by means of which the positions of the moon could be observed in other parts of her orbit, found that they sometimes agreed, but were more frequently at variance with the calculated places; the greatest amount of error always taking place at quadrature and vanishing altogether at syzygy.

What must, however, have been a source of great perplexity to Ptolemy, when he attempted to investigate the law of this new irregularity, was to find that it did not return in every quadrature,—in some quadratures it totally disappeared, and in others amounted to 2° 39′, which was its maximum value.

By dint of careful comparison of observations, he found that the value of this second inequality in quadrature was always proportional to that of the first in the same place, and was additive or subtractive according as the first was so; and thus, when the first inequality in quadrature was at its maximum or 5° 1′, the second increased it to 7° 40′, which was the case when the apse line happened to be in syzygy at the same time.*

* It would seem as if Hipparchus had felt the necessity for some further modification of his first hypothesis, though he was unable to determine it; for there is an observation made by him on the moon in the position here specified when the error of his tables would be greatest; and at a time also when she

But if the apse line was in quadrature at the same time as the moon, the second inequality vanished as well as the first.

The mean value of the two inequalities combined was therefore fixed at 6° 20½′.

121. To represent this new inequality, which was subsequently called the *Evection*, Ptolemy imagined an excentric in the circumference of which the centre of an epicycle moved while the moon moved in the circumference of the epicycle.

The centre of the excentric and of the epicycle he supposed in syzygy at the same time, and both on the same side of the earth.

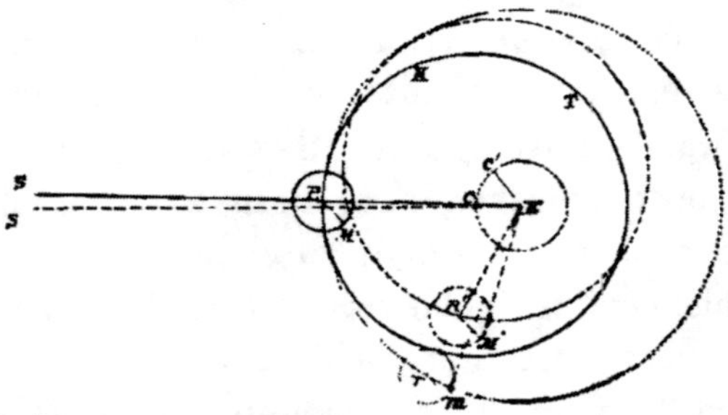

Thus, if $E$ represent the earth,
$S$ .................. sun,
$M$ .................. moon,
$c$ the centre of the excentric $RKT$ in syzygy,
$R$, the centre of the epicycle, would also be in syzygy.

Now conceive $c$, the centre of the excentric, to describe a small circle about $E$ in a retrograde direction $cc'$, while $R$, the centre of the epicycle, moves in the opposite direction,

was in the nonagesimal, so that any error of longitude, arising from her yet uncertain parallax, would be avoided. Ptolemy, who records the observation, employs it to calculate the evection, and obtains a result agreeing with that of his own observations. (See Delambre, *Ast. Ancienne.*)

in such a manner that each of the angles $S'Ec'$, $S'ER'$ may be equal to the synodical motion of the moon, that is, her mean angular motion from the sun; $SES'$ being the motion of the sun in the same time.

Now we have seen, Art. (116), that the first inequality was accounted for by supposing the epicycle $RM$ to move into the position $rm$, $r$ and $R$ being at the same distance from $E$, and $rm$ parallel to $RM$,* the first inequality being the angle $rEm$. But when the centre of the epicycle is at $R'$, and $R'M'$ is parallel to $rm$, the inequality becomes $R'EM'$, and we have a second correction or inequality $mEM'$.

122. That this hypothesis will account for the phenomena observed by Ptolemy, will be readily understood.

At syzygies, whether conjunction or opposition, the centres of the excentric and epicycle are in one line with the earth and on the same side of it; the points $r$ and $R'$ coincide, as also $m$ and $M'$. Hence $mEM' = 0$.

At quadratures (figs. 1 and 2) $c'$ and $R'$ are in a straight line on opposite sides of the earth, and therefore $R'$ and $r$ at

Fig. 1.

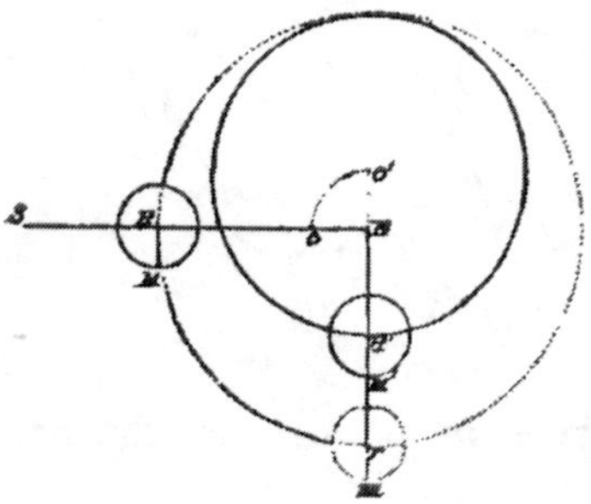

Fig. 2.

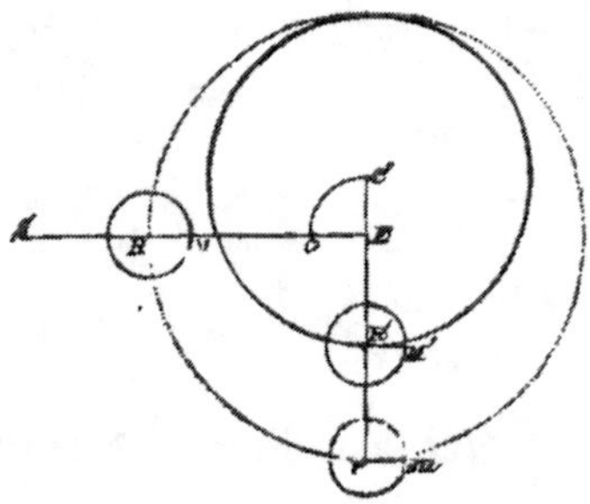

their furthest distance. If, however, $M'$ and $m$ be at the same time in this line, or, in other words, if the apse line

* For simplicity we leave out of consideration the motion of the apse.

be in quadratures (fig. 1), the angle $mEM'$ will still be zero, or there will be no error in the longitude. But, if the apse line is in syzygy (fig. 2), the angle $mEM'$ attains its greatest value.*

Ptolemy, as we have said, found this greatest value to be 2° 39′, the angle $mEr$ being then 5° 1′.

123. Copernicus (A.D. 1543), having seen that Ptolemy's hypothesis gave distances totally at variance with the observations on the changes of apparent diameter,† made another and a simpler one which accounted equally well for the inequality in longtitude, and was at the same time more correct in its representation of the distances.

Let $E$ be the earth, $OD$ an epicycle whose centre $C$ describes the circle $C'\ CC''$ about $E$ with the moon's mean angular velocity.

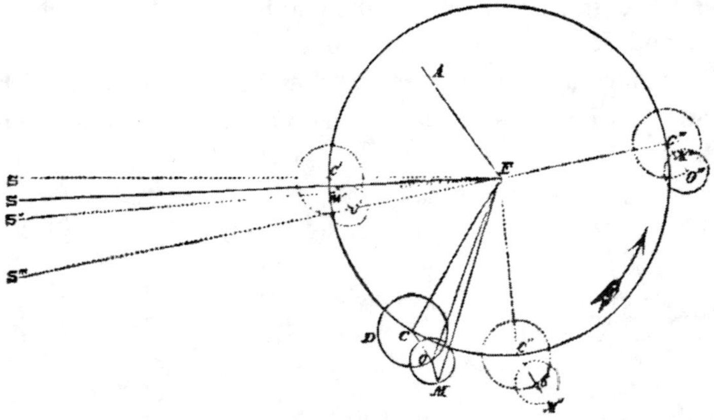


Let $CO$, a radius of this epicycle, be parallel to the apse

---

* If Ptolemy had used the hypothesis of an excentric instead of an epicycle for the first inequality of the moon, an epicycle would have represented the second inequality more simply than his method did. Dr. Whewell's *History of the Inductive Sciences*, vol. I., p. 230.

† See Delambre, *Ast. Moderne*, vol. I., p. 116. Whewell's *History of Inductive Sciences*, vol. I., p. 395.

line $EA$, and about $O$ as centre let a second small epicycle be described, the radii $CO$ and $OM$ be so taken that

$$\frac{CO - OM}{CE} = \sin 5^\circ\ 1', \text{ and } \frac{CO + OM}{CE} = \sin 7^\circ\ 40'.$$

The radius $OM$ must now be made to revolve from the radius $OC$ twice as rapidly as $EC$ moves from $ES$, so that the angle $COM$ may be always double of the angle $CES$.

From this construction, it follows that in syzygies the angle $CES$ being 0° or 180°, the angle $COM$ is 0° or 360°; and therefore $C$ and $M$ are at their nearest distances, as in the positions $C'$ and $C'''$ in the figure. Then $CM = CO - OM$, and the angle $CEM$ will range between 0° and 5° 1', the greatest value being attained when the apse line is in quadrature.

When the moon is in quadrature $CES = 90°$ or 270°, aud, therefore, $CGM = 180°$ or 540° and $C$ and $M$ are at their greatest distance apart, as in the position $C''$; then, $CM = CO + OM$, and the angle $CEM$ will range between 0° and 7° 40', the former value when the apse line is itself in quadrature, and the latter when it is in syzygy.

124. Thus the results attained by Ptolemy's construction are, as far as the longitudes at syzygies and quadratures are concerned, as well represented by that of Copernicus; and the variations in the distances of the moon will be far more exact, the least apparent diameter being 28' 45" and the greatest 37' 33"; whereas, Ptolemy's would make the greatest diameter 1°.*

The values which modern observations give vary between 28' 48" and 33' 32".

---

* Delambre, *Ast. Moderne.*

125. It will not now be difficult to shew that the introduction of this small epicycle corresponds with that of the term $\frac{15}{4}me\sin\{(2-2m-c)pt-2\beta+\alpha\}$ in our value of $\theta$.

For, referring to the preceding figure, we have

$$OEM=\sin OEM=\frac{OM}{OE}\sin OME$$
$$=\frac{OM}{OE}\sin(COM-AEM)$$
$$=\frac{OM}{OE}\sin(2\,.\,SEC-AEM)$$
$$=\frac{OM}{OE}\sin\{2\,(\text{moon's } \textit{mean} \text{ long.} - \text{sun's long.})$$
$$- (\text{moon's } \textit{true} \text{ long.} - \text{long. of apse})\},$$

and $OEM$ being a small angle whose maximum is $1°\ 19\frac{1}{2}'$, we may write moon's mean longitude instead of the true in the argument, and also $EC$ for $OE$; therefore,

$$OEM=\frac{OM}{EC}\sin\{2\,(\text{moon's mean longitude} - \text{sun's longitude})$$
$$- (\text{moon's mean longitude} - \text{longitude of apse})\}$$
$$=79\tfrac{1}{2}'\sin[2\,\{pt-(mpt+\beta)\}-\{pt-\overline{(1-c)pt+\alpha}\}]$$
$$=4770''\sin\{(2-2m-c)pt-2\beta+\alpha\}.$$

The value of the coefficient is from modern observations found to be 4589·61″.

126. In Art. (77), we have considered the effect of this second inequality in another light, not simply as a small quantity additional to the first or elliptic inequality, but as forming a part of this first; and therefore, modifying and constantly altering the excentricity and the uniform progression of the apse line.

Boulliaud (A. D. 1645), by whom the term Evection was first applied to the second inequality, seems to hint at something of this kind in the rather obscure explanations of *his* lunar hypothesis, which, never having been accepted, it would be useless to give an account of it.*

In Ptolemy's theory, Art. (121), the evection was the result of an *apparent* increase of the first lunar epicycle caused by its approaching the earth at quadratures; but, in this second method, it is the result of an *actual* change in the elements of the elliptic orbit.

D'Arzachel, an Arabian astronomer, who observed in Spain about the year 1080, seems to have discovered the unequal motion of the apsides, but his discovery must have been lost sight of, for Horrocks, about 1640, re-discovered it 'in consequence of his attentive observations of the lunar 'diameter: he found that when the distance of the sun from 'the moon's apogee was about 45° or 225°, the apogee was 'more advanced by 25° than when that distance was about '135° or 315°. The apsides, therefore, of the moon's orbit 'were sometimes progressive and sometimes regressive, and 'required an equation of 12° 30′, sometimes additive to their 'mean place and sometimes subtractive from it.'†

Horrocks also made the excentricity variable between the limits ·06686 and ·04362.

The combination of these two suppositions was a means of avoiding the introduction of Ptolemy's excentric or the second epicycle of Copernicus: their joint effect constitutes the evection.

---

* Après avoir établi les mouvements et les époques de la lune, Boulliaud revient à l'explication de l'évection ou de la seconde inégalité. Si sa théorie n'a pas fait fortune, le nom du moins est resté. 'En même temps qne la lune avance sur son 'cône autour de la terre, *tout le système de la lune est déplacé*; la terre emportant la 'lune, rejette loin d'elle l'apogée, et rapproche d'autant le périgée; mais cette 'évection à des bornes fixées.' Delambre, *Hist. de l' Ast. Mod.*, tom. II. p. 157.

† Small's *Astronomical Discoveries of Kepler*, p. 307.

*Variation.*

127. After the discovery of the evection by Ptolemy, a period of fourteen centuries elapsed before any further addition was made to our knowledge of the moon's motions. Hipparchus's hypothesis was found sufficient for eclipses, and when corrected by Ptolemy's discovery, the agreement between the calculated and observed places was found to extend also to quadratures; any slight discrepancy being attributed to errors of observation or to the imperfection of instruments.

But when Tycho Brahé (A. D. 1580) with superior instruments extended the range of his observations to all intermediate points, he found that another inequality manifested itself. Having computed the places of the moon for different parts of her orbit and compared them with observation, he perceived that she was always in advance of her computed place from syzygy to quadrature, and behind it from quadrature to syzygy; the maximum of this *variation* taking place in the octants, that is, in the points equally distant from syzygy and quadrature. The moon's velocity therefore, so far as this inequality was concerned, was greatest at new and full moon, and least at the first and third quarters.*

* 'It appears that Mohammed-Aboul-Wefa-al-Bouzdjani, an Arabian astronomer of the tenth century, who resided at Cairo, and observed at Bagdad in 975, 'discovered a third inequality of the moon, in addition to the two expounded by 'Ptolemy, the equation of the centre and the evection. This third inequality, the 'variation, is usually supposed to have been discovered by Tycho Brahé, six 'centures later......In an almagest of Aboul-Wefa, a part of which exists in 'the Royal Library at Paris, after describing the two inequalities of the moon, 'he has a Section IX., "Of the third anomaly of the moon called *Muhazat* or '*Prosneusis*"......But this discovery of Aboul-Wefa appears to have excited 'no notice among his contemporaries and followers; at least it had been long 'quite forgotten, when Tycho Brahé re-discovered the same lunar inequality.' Whewell's *Hist. of Inductive Sciences*, vol. I. p. 243.

Tycho fixed the maximum of this inequality at 40′ 30″. The value which results from modern observations is 39′ 30″.

128. We have already two epicycles, or one epicycle and an excentric, to explain the first two inequalities: by the introduction of another epicycle or excentric, the variation also might have been brought into the system; but Tycho adopted a different method:* like Ptolemy, he employed an excentric for the evection, but for the first or elliptic inequality he employed a couple of epicycles, and this complicated combination, which it is needless further to describe, represented the change of distance better than Ptolemy's.

To introduce the *variation*, he imagined the centre of the larger epicycle to librate backwards and forwards on the excentric, to an extent of 40½′ on each side of its mean position; this mean place itself advancing uniformly along the excentric with the moon's mean motion in anomaly; and the libration was so adjusted, that the moon was in her mean place at syzygy and quadrature, and at her furthest distance from it in the octants, the period of a complete libration being half a synodical revolution.

### *Annual Equation.*

129. Tycho Brahé was also the discoverer of the fourth inequality, called the annual equation. This was connected with the anomalistic motion of the sun, and did not, like the previous inequalities, depend on the position of the moon in her orbit.

---

* For a full description of Tycho's hypothesis, see Delambre, *Hist. de l'Ast. Mod.* tom. I. p. 162, and *An Account of the Astronomical Discoveries of Kepler* by Robert Small, p. 135.

Having calculated the position of the moon corresponding to any given time, he found that the observed place was behind her computed one while the sun moved from perigee to apogee, and before it in the other half year.

Tycho did not state this distinctly, but he made a correction which, though wrong in quantity and applied in an indirect manner, shewed that he had seen the necessity and understood the law of this inequality.

He did not try to represent it by any new excentric or epicycle, but he increased by (8m. 13s.) sin (*sun's anomaly*) the time which had served to calculate the moon's place;* thus assuming that the true place, after that interval, would agree with the calculated one. Now, as the moon moves through 4′ 30″ in 8m. 13s., it is clear that adding (8m. 13s.) sin (*sun's anomaly*) to the time is the same thing as subtracting (4′ 30″) sin (*sun's anomaly*) from the calculated longitude, which was therefore the correction virtually introduced by Tycho.† Modern observations shew the coefficient to be 11′ 9″.

We have seen, Art. (82), how this inequality may be inferred from our equations.

### *Reduction.*

130. The next inequality in longitude which we have to consider is not an inequality in the same sense as the foregoing; that is, it does not arise from any irregularity in the motion of the moon herself in her orbit, but simply because that orbit is not in the same plane as that in which the

---

* That is, the equation of time which he used for the moon differed by that quantity from that used for the sun.

† Horrocks (1639) made the correction in the same manner as Tycho, but so increased it that the corresponding coefficient was 11′ 51″ instead of 4′ 30″. Flamsteed was the first to apply the correction to the longtitude instead of the time.

longitudes are reckoned, so that even a regular motion in the one would be necessarily irregular when referred to the other. Thus if $NMn$ be the moon's orbit and $♈Nm$ the ecliptic, and if $M$ the moon be referred to the ecliptic by the great circle $Mm$ perpendicular to it, then $MN$ and $mN$ are 0°, 90°, 180°, 270°, and 360° simultaneously, but they differ for all intermediate values: the difference between them is called the *reduction*.

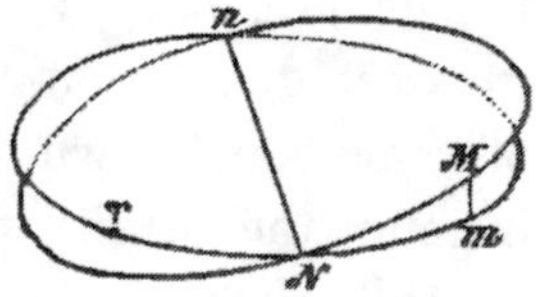


The difference between the longitude of the node and that of the moon in her orbit being known, that is the side $NM$ of the right-angled spherical triangle $NMm$, and also the angle $N$ the inclination of the two orbits, the side $Nm$ may be calculated by the rules of spherical trigonometry, and the difference between it and $NM$, applied with a proper sign to the longitude in the orbit, gives the longitude in the ecliptic.

Tycho was the first to make a table of the reduction instead of calculating the spherical triangle. His formula was

$$\text{reduction} = \tan^2 \tfrac{1}{2} I \sin 2L - \tfrac{1}{2} \tan^4 \tfrac{1}{2} I \sin 4L,$$

where $I$ is the inclination of the orbit and $L$ the longitude of the moon diminished by that of the node.

The first term corresponds with the term $-\frac{1}{4}k^2 \sin 2(gpt - \gamma)$ of the expression for $\theta$.

### *Latitude of the Moon.*

131. That the moon's orbit is inclined to the ecliptic was known to the earliest astronomers, from the non-recurrence of eclipses at every new and full moon; and it was also known, sinces the eclipses did not always take place in the same parts of the heavens, that the line of nodes represented

by $Nn$, in the preceding figure, has a retrograde motion on the ecliptic, $N$ moving towards ♈.

Hipparchus fixed the inclination of the moon's orbit to the ecliptic at 5°, which value he obtained by observing the greatest distance at which she passes to the north or south of some star known to be in or very near the ecliptic, as for instance the bright star Regulus; and by comparing the recorded eclipses from the times of the Chaldean astronomers down to his own, he found that the line of nodes goes round the ecliptic in a retrograde direction in about $18\frac{2}{3}$ years.

This result is indicated in our expression for the value of the latitude by the term $k \sin (g\theta - \gamma)$, as we have shewn Art. (85).

132. Tycho Brahé further discovered that the inclination of the lunar orbit to the ecliptic was not a constant quantity of 5° as Hipparchus had supposed, but that it had a mean value of 5° 8′, and ranged through 9′ 30″ on each side of this, the least inclination 4° $58\frac{1}{2}$′ occurring when the node was in quadrature, and the greatest 5° $17\frac{1}{2}$′ being attained when the node was in syzygy.*

He also found that the retrograde motion of the node was not uniform: the mean and the true positions agreed very

---

* Ebn Jounis, an Arabian astronomer (died A.D. 1008), whose works were translated about 50 years since by Mons. Sedillot, states that the inclination of the moon's orbit had been often observed by Aboul-Hassan-Aly-ben-Amajour about the year 918, and that the results he had obtained were generally greater than the 5° of Hipparchus, but that they *varied considerably*.

Ebn Jounis adds, however, that he himself had observed the inclination several times and found it 5° 3′, which leads us to infer that he always observed in similar circumstances, for otherwise a variation of nearly 23′ could scarcely have escaped him. See Delambre, *Hist. de l' Ast. du Moyen Age*, p. 139.

The mean value of the inclination is 5° 8′ 55·46″,—the extreme values are 4° 57′ 22″ and 5° 20′ 6″.

The mean daily motion of the line of nodes is 3′ 10·64″, or one revolution in 6793·29 days, or 18 y. 218 d. 21 h. 22m. 46 s.

well when they were in syzygy or quadrature, but they were 1° 46′ apart in the octants.

By referring to Art. (87) we shall see that these corrections, introduced by Tycho Brahé, correspond to the second term of our expression for *s*.

Since Hipparchus could observe the moon with accuracy only in the eclipses, at which time the node is in or near syzygy, we see why he was unable to detect the want of uniformity in the motion of the node.

133. To represent these changes in the position of the moon's orbit, Tycho made the following hypothesis.

Let *ENF* be the ecliptic, *K* its pole, *BAC* a small circle,

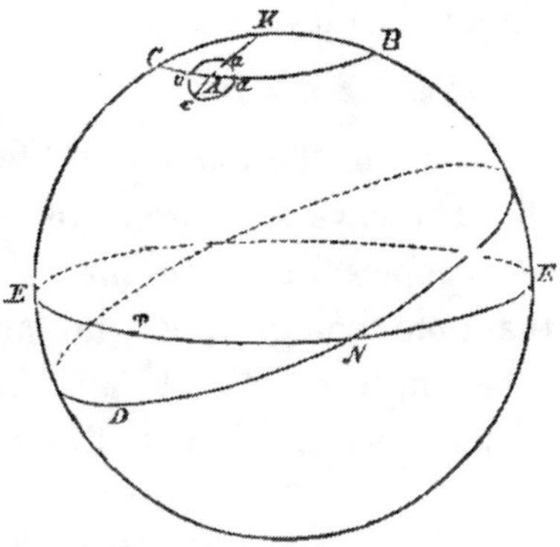


having also *K* for pole and at a distance from it equal to 5° 8′. Then, if we suppose *A* the pole of the moon's orbit to move uniformly in the small circle and in the direction *BAC*, the node *N*, which is at 90° from both *A* and *K*, will retrograde uniformly on the ecliptic, and the inclination of the two orbits will be constant and equal to *AK*.

But instead of supposing the pole of the moon's orbit to be at *A*, let a small circle *abcd* be described with *A* as pole and a radius of 9′ 30″; and suppose the pole of the moon's orbit to describe this small circle with double the velocity of the node in its synodical revolution which is accomplished in about 346 days, in such a manner that when the node is in quadrature the pole may be at $a$, the nearest point to $K$, and at $c$ the most distant point when the node comes to syzygy, at $d$ in the first and third octants, and at $b$ in the second and fourth, so as to describe the small circle in about 173 days, the centre $A$ of the small circle retrograding meanwhile with its uniform motion.

By this method of representing the motion, we see that

| | | |
|---|---|---|
| when node is in quadrature | the inclination of the orbit is | $Ka = 5°8' - 9\frac{1}{2}' = 4°58\frac{1}{2}'$, |
| .................. syzygy | | $Kc = 5°8' + 9\frac{1}{2}' = 5°17\frac{1}{2}'$, |

while at the octants it has its mean value

$$Kb = Kd = KA = 5°\ 8'.$$

Again, with respect to the motion of the node, since $N$ is the pole of $KaAc$, it follows that when in syzygy and quadrature, the node occupies its mean place; in the first and third octants, the pole being at $d$, the node is before its mean place by the angle $dKA = (9'\ 30'')\ \text{cosec}\, 5°\ 8' = 1°\ 46'$, nearly, and it is as much behind in the second and fourth octants.

So that the whole motion of the node, and the correction which Tycho had discovered, were properly represented by this hypothesis, which is exactly similar to that which Copernicus had imagined to explain the precession of the Equinox.

# SELECTION OF EXAMINATION QUESTIONS

## FROM COLLEGE AND SENATE-HOUSE PAPERS AND FROM THE MODERATORSHIP AND FELLOWSHIP EXAMINATIONS AT TRINITY COLLEGE, DUBLIN.

1. Define the plane of the ecliptic and prove that, as seen from the earth, $\sin\lambda = \frac{1}{32400}\sin\lambda'$ nearly; $\lambda$ being the latitude of the sun and $\lambda'$ that of the moon.

2. Obtain the differential equation of the moon's radius vector.

3. What is the principle on which the successive approximations to the moon's path are obtained?

Compare the coefficients of the principal term in the central disturbing force with the principal term in the central force.

4. If in solving the equation

$$\frac{d^2u}{d\theta^2} + u = \frac{P}{h^2u^2} - \frac{T}{h^2u^3}\frac{du}{d\theta} - 2\left(\frac{d^2u}{d\theta^2} + u\right)\int\frac{T}{h^2u^3}\,d\theta$$

we take as our first approximation $u = a\,\{1 + e\cos(\theta - \alpha)\}$, we obtain at our higher approximations terms containing $\theta$ in their coefficient. How is this defect avoided by taking $u = a\,\{1 + e\cos(c\theta - \alpha)\}$ as our first approximation?

Upon what principle do we approximate to the value of $c$?

5. Shew that in the expansion of $\frac{T}{h^2u^3}$ there will be a term $-\frac{15}{4}m^2e^2\sin\{(2 - 2m - 2c)\,\theta - 2\beta + 2\alpha\}$ which would rise to the second order in the longitude.

Shew further that the term with the same argument does not appear *actually* in the longitude—assuming

$$u = a\,[1 + e\cos(c\theta - \alpha) + \ldots + \tfrac{15}{8}me\cos\{(2 - 2m - c)\,\theta - 2\beta + \alpha\} + \ldots + \tfrac{15}{4}me^2\cos\{(2 - 2m - 2c)\,\theta - 2\beta + 2\alpha\} + \ldots].$$

6. The longitude of the moon contains a term of the form

$$A \sin\{(2-2m-c)\,pt-2\beta+a\}$$

where $A$ is of the second order; find how it must have arisen and determine $A$.

7. Investigate the following expression for the moon's longitude as far as the second order, supposing the orbits of the moon relative to the earth, and of the earth relative to the sun, to be originally circles in the plane of the ecliptic,

$$\theta = pt + \tfrac{11}{8}m^2 \sin\{(2-2m)\,pt-2\beta\}.$$

8. Suppose that whatever constants are involved in the arguments $\theta$, $\phi$, $\psi$, &c. in the equation

$$L = A + B\sin\theta + C\sin\phi + D\sin\psi + \&c.$$

are perfectly known, and that observation gives the numerical values of a considerable number of longitudes $L_1$, $L_2$, $L_3$, &c., corresponding to the known angles $\theta_1, \phi_1, \psi_1$, &c., $\theta_2, \phi_2, \psi_2$, &c., $\theta_3$, $\phi_3$, $\psi_3$, &c., &c., shew how the constants $A$, $B$, $C$, &c. may be numerically determined in the two distinct cases where the observations at our disposal *are* and *are not* unlimited in number.

9. Explain how this process fails and may be modified when two of the terms in question $B\sin\theta$, $C\sin\phi$ are nearly synchronous in their periods.

10. Shew that the evection in longitude, viz.

$$\tfrac{15}{4}me \sin\{(2-2m-c)\,pt-2\beta+\alpha\},$$

may be represented as the joint effect of certain periodic changes in the excentricity of the lunar orbit and in the mean longitude of its apse.

11. Assuming the usual notation of the Lunar Theory, explain the physical meaning of the following equations:

$$s = k\sin(\theta-\gamma),$$
$$s = k\sin(g\theta-\gamma),$$
$$s = k\sin(g\theta-\gamma) + \tfrac{3}{8}mk\sin\{(2-2m-g)\,\theta-2\beta+\gamma\}.$$

12. Explain the effect of the term $m^2a\cos\{(2-2m)\,\theta-2\beta)\}$ in the moon's radius vector; find the number of days in

the period of the resulting inequality and the ratio of the axes of the oval orbit.

13. Investigate the effect of the annual equation on the length of the lunar month.

Assuming $m = \cdot 075$, $e' = \cdot 017$, $\sin 13^\circ 30' = \cdot 23345$, and the mean period of a sidereal revolution of the moon = 27d. 7h. 43m. 12s., find the difference between a winter and a summer month.

14. Shew that the moon's orbit in space is always concave towards the sun.

Assuming that the earth describes a circle about the sun, and the moon a circle about the earth and in the same plane with the earth's orbit, compare the curvatures of the moon's orbit in space at her perihelion and aphelion.

15. Prove that if we go to the third order of approximation, the motion of the moon's apse in one revolution of the moon equals $\frac{3}{4}m^2\left(1 - \frac{75}{8}m\right) 360^\circ$.

16. The expressions for $\frac{P}{h^2u^2}$ and $\frac{T}{h^2u^3}$ contain terms of the form $3A\cos(\theta - \theta')$ and $A\sin(\theta - \theta')$ respectively; compute the terms resulting from them in the values of $u$ and $\theta$.

17. Copernicus represented the evection by an epicycle superposed on the epicycle which represented the elliptic inequality. From the construction given by him shew geometrically that the evection vanishes when the moon's mean elongation from the sun is half her true anomaly, and, in general, varies nearly as the sine of twice the difference of these angles.

18. The secular change of inclination of the actual ecliptic to a fixed ecliptic being $\alpha''$ annually, shew that this will give rise to an inequality in the moon's latitude whose type is $-\frac{n}{2\pi}\alpha''\cos\phi$, nearly; $n$ being the number of years in which the moon's nodes make a complete revolution, and $\phi$ the difference of longitude of the moon and of the ascending node of the actual ecliptic.

## BY THE SAME AUTHOR.

---

I.

A TREATISE ON ASTRONOMY, for the use of Colleges and Schools. 8vo. cloth, 12*s*. 6*d*.

MACMILLAN AND CO.

II.

A CHART AND DIAGRAM for Facilitating Great-Circle Sailing, 3*s*. 6*d*.

With a Paper from Part II., Vol. X. of the *Cambridge Philosophical Transactions*.

III.

A TIME-AZIMUTH DIAGRAM for Determining the Time-Azimuth without Calculation. 3*s*. 6*d*.

With a Pamphlet explaining its use and construction.

*These Diagrams have been engraved by the Hydrographic Office, Admiralty. Published by J. D. Potter, 31, Poultry, London, Agent for the Admiralty Charts.*